Corrugated Paper
Packaging Design

Contents

038 Printing on Corrugated Cardboard

042 Cutting and Creasing of Corrugated Cardboard

044 Development Trend of Corrugated Packaging

050 Cases

254 Index

Preface

We belong to a generation that witnessed the decline of plastic and the death of modernity at the end of the 90s, and is currently developing in the discovery of new forms of communication, the emergence of emotional advertising and the shift of society as a whole, and therefore businesses, towards a paradigm of sustainability through recycling.

This places us in a context of change, but especially where design is concerned, when have things been otherwise?

The re-industrialization of creativity.

As I see it, albeit in a different way and with caveats that I will deal with later, the market is currently undergoing a second industrial revolution, in which corrugated cardboard plays a major role. For aesthetic evidence, we only need to pick up a catalogue of design furniture, which these days finds its inspiration in rusty iron factories; or watch one of the movies featuring that incredibly aesthetic, futuristic and romantic new trend known as steampunk, with its dreams of a return to the age of steam and industry for the sake of a more humane society, free of the excesses of dehumanized technology we see today.

Awareness, ecology, sustainability, recycling, creativity... are all now synonymous with one thing: cardboard. Art directors search excitedly for experts in paper and cardboard craft with the ability to use them in producing accurate representations of reality. Nothing less than a piece of engineering in which, it would seem, the emphasis is now on detail and taking the time to create small works of art, a slight vestige of the art & craft movement, now at the service of multinationalisation and globalisation, where the leading brands reproduce their logos in cardboard... just as

Mr. William Morris would have always wished.

It is abundantly obvious, and here this book can serve as an example, that corrugated cardboard is right now on the crest of the wave in design worldwide. Here, I think it is important to take a stand for design, in whatever form, as a universal yardstick, mirror, and accurate means of analysis of current aesthetic and intellectual trends.

Until now, corrugated cardboard, with its brown colour and simplicity of form, has systematically appeared before us whenever we have approached the world of industry, where factories, great and small, needed a practical packaging solution for their products. Design, colour, and creativity gave way to practicality and economy. Thus, the production of cardboard and its transformation into boxes became standardised to meet with requirements, eventually giving way to easily affordable standard sizes and formats that did not require complex dies. It has to be said that this economic and logistic advantage did not, at the time, in my opinion, receive the acknowledgement it deserved in ecological terms, whereas today, thanks to society's change of mentality, this is virtually its greatest value.

It could be said, therefore, that its use and function within this small industrial and practical ecosystem, were perceived as commonplace by us members of society.

This, for me, marks a major turning point. In the words of the philosopher Trías 'The commonplace, properly understood, can, and has demonstrated as much, become art, thought, beauty…'. Its value, over and above its practicality, was there and only needed to be brought to light. If analysed, this thought could reveal many keys. To quote the second epistle to the Corinthians 2:4:18:

...while we look not at the things which are seen, but at the things which are not seen: for the things which are seen are temporal; but the things which are not seen are eternal...

The value of what is hidden, but nevertheless exists, is eternal. Of this we, communication professionals, are well aware. When you are asked to create an advertising or branding campaign, the first thing you have to highlight and attempt to identify is that value in a label or advertising jingle which, despite being obvious, is nevertheless hidden from the consumer at first sight.

Nowadays, cardboard is everywhere, it has quite naturally worked its way into our lives, as if it has always been there, like an ever-present spirit that has only now made itself visible.

In the world of furniture and interior design, for example, cardboard is becoming established as a synonym of simplicity and practicality, still maintaining its intrinsic virtues. At the same time, however, it is also a synonym of elegance, trendiness, space, design... Major department stores use it in their window displays to represent animals and other elements of nature. These silhouettes and sometimes more complex 3D models convey the essence of the everyday objects they represent, transporting us into a world and a concept in which what matters is the origin, the return to a slower pace, to what is universal, to nature, and in contrast, the ephemeral nature of mankind.

We are also used to seeing how a couple of pieces of cardboard placed one on top of the other can become educational materials, tables and chairs for children, a celebration of the pastime or creativity in play, or the basis for a flight of the imagination. As Matisse put it, art should be 'rather like a good armchair which provides relaxation from physical fatigue', which is something that all of today's designers must keep in mind, or at least they should if they want their client firms to keep abreast of the times and make communication.

I have always thought that any design (and, all the more, advertising) which fails to communicate is like talent with no underlying effort, and in the case of cardboard, its value and ability to communicate have been with us since its origins. Furthermore, let us not kid ourselves, nowadays, cardboard sells.

In packaging, the concept changes little. In the information-saturated world of today, it is becoming increasingly difficult to find room on packages for all the information the customer requires and every millimetre of permitted printing space counts. In this respect, corrugated cardboard has already said a great deal. Its innate characteristics, its resistance to being printed in bright colours and its roughness for achieving fine detail have all contributed to the simplicity of design it tends to feature.

Unlike other materials, cardboard has greatly simplified the message in the food sector, where a surfeit of colours and lacquers used to be the norm and type of product took centre stage. On the whole, large stains and print are giving way to cardboard, which, as I am trying to explain, speaks its own language to the consumer. Right now, structure is taking over from appearance; the material is upstaging the image, while keeping its own identity intact.

Through concern for sustainability and the protection of nature this ancestral spirit has become visible in our civic social and ethical objectives. An ecological reindustrialization is underway. And I use the term reindustrialization because even today, due to the industrial heritage of cardboard, the complexity of producing it in small quantities and the unwillingness of the industrial giants to cater for the creative world, small-scale supplies of corrugated packaging are still hard to obtain and we are left at the mercy of the major die-cutting firms, with all the expense this entails.

Nevertheless, this has certainly not stopped the big companies from making the most of all the advantages it has to offer. Right now, its sustainability offers huge potential that we in the world of advertising know how to put to use, and sometimes, it has to be said, even overuse.

I would even go so far as to describe it as universal; but in a positive sense, because the term can have its negative connotations. The homeless use cardboard to keep out the cold when sleeping rough on the streets; beggars present their pitiful requests printed on pieces of milk carton packages. At the same time, however, behind this revolution of industrial creativity, marketing departments want to use it to dress up their abandonment of plastic and shiny stucco; while perfume companies raise the price of their gift sets after packaging them in corrugated cardboard to fabricate the illusion of an adventure like those of yesteryear.

For an example, take some of the latest international campaigns that have passed through our studio, where the overriding objective is to create a universal message that works just as well in the Arab world as in South America. In two cases out of three, sooner or later in the meeting, the client will let slip the words recycled, cardboard or ecological.

And there's no denying the fact: the values contained in the material that forms the topic of this book are there for all cultures. Nor can it be denied that the endless categories and corporate social responsibility logos that appear on packages can drive consumers crazy, so they can be forgiven if they tend to generalize, which they often do, and confuse corrugated cardboard packaging with the idea of own brand, inferior quality, or lack of resources.

Here, and may this book serve as an example, the responsibility lies with us designers and our ability to make proper use of a material whose honesty, beauty, strength of identity and wealth of possibilities are as great as our individual creativity is able to reveal.

Germán Úcar
estudioVACA's founding partner

Introduction |

As technology progresses and people's awareness of environmental protection increases, corrugated paper has witnessed a leap in its application in packaging. It is a green packing material of light weight, high strength; it is easy to mould, fold, store and transport; also easy for recycling and reuse; it features good printability and low cost, etc. It can be found in every sector in daily life, covering the consumer package and transport package of food, beverage, household products, electronic devices, industrial equipments, and stationary.

Corrugated fiberboard is a flat packaging material of fluted corrugated sheet and linerboards, often referred to as 'corrugated cardboard'. Regular corrugated cardboard, honeycomb cardboard and micro-corrugated cardboard are three most common types of corrugated material.

Common Forms of Corrugated Cardboard |

Corrugated cardboard is also called corrugating cardboard. The ridges and grooves are like a row of arched doors, parallel to and supporting each other to form a triangular structure which has good mechanical strength. Its surface can bear certain degree of pressure and has good elasticity and buffering performance; it can be moulded into liner or container of all shapes and sizes, more convenient to produce than plastic cushioning materials; it is little affected by temperature, has good shading performance and does not degenerate in sunlight; it is also little affected by humidity (See figure 1).

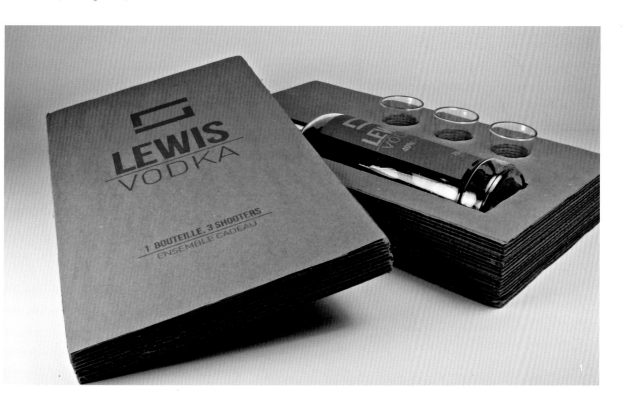

Corrugated cardboard package designed by Elyse Levass

Honeycomb cardboard is a new material, featuring light weight, high degree of utilisation and lower total cost; high strength, smooth surface, less likeliness of deformation; easy for decoration, large variety of products, wide application; biodegradable, recyclable, conform to environmental requirements; shock absorbing, sound absorbing, heat insulating, durable. Honeycomb cardboard can be made into trays, large boxes and cushioning. It is a green packaging material to be used instead of bamboo/wood trays, bamboo/wood boxes, heavy corrupted boxes and EPS foam plastic cushioning (See figure 2).

Honeycomb cardboard package by O Zone design

Micro-corrugated cardboard is a new type of corrugated paper gaining popularity over the years. As many consumer goods are presented in small or personalised packages, transported in small quantities, it is necessary that the packaging can provide protective properties and also bring new market potential with less weight. Micro-corrugated packaging can increase the pressure resistance, reduce the weight of packaging, and provide satisfying printability according to related research. Therefore designers are able to use different corrugated cardboard for different products, considering the requirement of different weight, type and category. The advantages of corrugated cardboard are getting more and more attention, as it has been widely used in the packaging of digital products, household appliances, food, medical equipments, etc. Micro-corrugated is always the choice of priority in mobile phone packaging (See figure 3).

Nick Holt's micro-corrugated package design

Advantages of Corrugated Packaging |

Corrugated packaging features light weight and excellent structural properties. With an arched internal structure, corrugated cardboard can provide satisfying shock absorption and mechanical performance. There are also many other protective properties including the protection against humidity, easy cooling and transport.

The change in specification and dimension is easy to achieve, and can easily adapt to packaging of all categories.

Easy to print, corrugated cardboard can meet the requirements of decoration or printing design for all kinds of cartons.

Corrugated packaging is strong and durable, therefore extra packaging is not needed during promotional display or transport. Besides, it is easy to pack and handle with mechanised and automated equipments, saving lots of resources.

Using environmental friendly raw materials, corrugated products are recyclable and reusable. They are made of wood fibres which can decompose in the natural environment. And with organised planting and logging, timber resource exhaustion can be avoided. Most abandoned corrugated cartons can be recycled and made into new products, contributing to environmental protection.

Slotted Carton

Slotted carton is a one piece carton with two sets of flaps on both ends, featuring less material consumption, better strength and high productivity. The width of the flat package is the same when it is closed and sealed, with the flaps meeting in the middle to form double top and bottom. It is less resource-efficient than two part cartons, thus usually used in the packaging for heavier items such as art crafts (See figure 4).

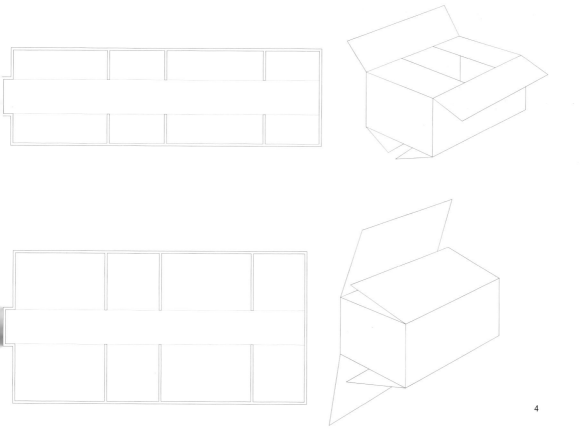

4

Full/Partial Telescope Carton

Telescope carton is a two piece carton consisting of a top and a bottom of different but matching sizes. It is used with the top telescoped over the bottom, usually in the packaging of soft and light items like textiles and printed fabrics. With the top and bottom fitting close to each other and tied together with packing tape, it can reduce the volume of commodities, save space and reduce the cost in transport (See figure 5).

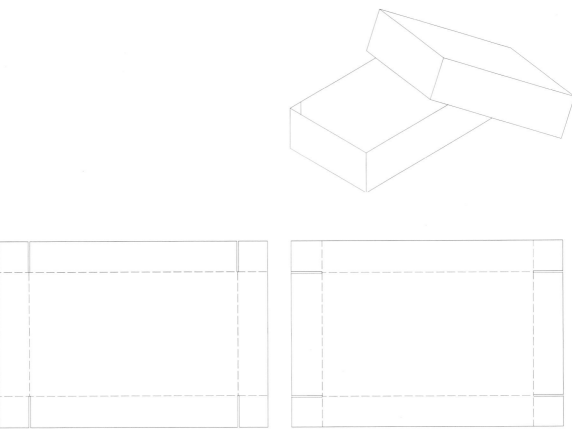

Half Slotted Carton

Half slotted carton is usually small, manufactured with a die-cut machine. It is often used in the packaging of soft items, e.g. knitwear, embroidery slippers, grass crafts as it features smooth edge and precise specification. The bottom needs no adhesive or sealing tape, adding convenience in use (See figure 6).

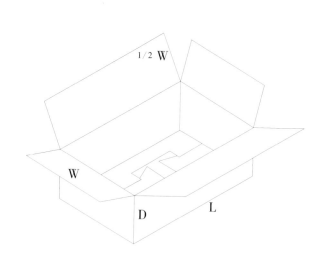

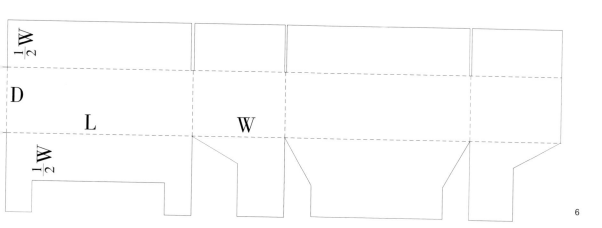

Full Overlap Carton

The structure of a full overlap carton is similar to that of wooden boxes. Stapling, gluing or tape sealing is needed before use to assembly the parts into a carton. With satisfying strength and weight-bearing performance, this kind of corrugated carton is widely used for the packaging of household items, stationery and food (See figure 7).

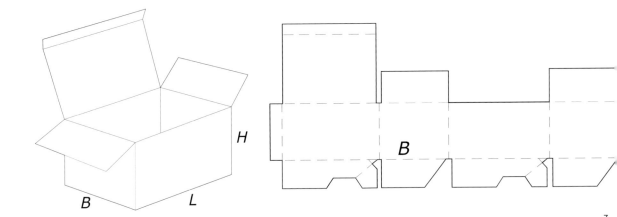

7

Micro-corrugated Carton

Micro-corrugated carton is often used as the packaging for smaller items. It is more creative than the traditional slotted and overlap cartons, as there are seemingly endless structural designs and uses. According to the requirements of specific products, a great variety of micro-corrugated cartons are available to present an attractive and sophisticated image (See figure 8).

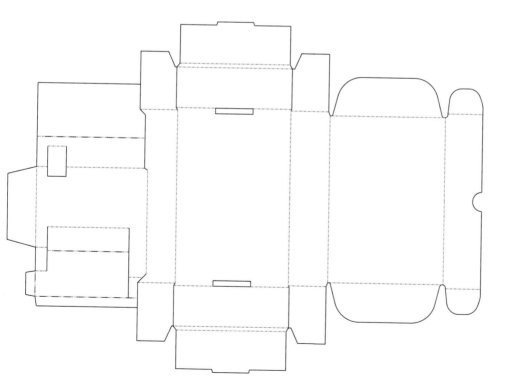

8

Guidelines in Corrugated Packaging Design |

Environmental friendly. A top reason for using corrugated packaging is its feature of being environmental friendly. It is therefore preferable to use non-polluting auxiliary materials in the design process, e.g. colouring and ink.

Resource efficient. To achieve the best protective performance with the least material consumption not only contributes to conserving resources, but reducing economic costs. It is a big challenge for the designer.

Good performance in promotion and display. This emphasises on the graphic design of the corrugated packaging. It needs to be aesthetic to attract the consumers and promote purchase (See figure 9).

The design of egg box by Anita Vaskó

Preparation

Decide the basic product characteristics and understand the client's requirement. A corrugated packaging design should begin with understanding the basic characteristics of the items involved, including the weight, size, centre of gravity, vulnerable parts, structural weak links and fragile degree, with fragile degree being the most important factor. It is also necessary to actively communicate with the clients and understand their needs and requirements.

Decide the method of packaging. Different packing methods are taken depending on the size of the products. Items of large volume such as washing machines, refrigerators, air conditioners are usually packed by hand; commodities of smaller volume like food and drinks are more feasible for mechanised and automated packing. As packing material of higher quality is required in mechanized packing, it is necessary that the corrugated cardboard is of precise and smooth standard.

Make sure of the transport and storage conditions. After a full understanding of the items to be packed, it is also necessary to know the transport and storage conditions. Transport can be carried out via water, air and land, while land transport consists of automobile transport and train transport. The specific factors to pay attention to during storage are temperature and humidity of the warehouse, space available in the warehouse as it decides height of the stack. These factors have a direct influence on the choice of the carton type and material.

Choosing Corrugated Material and Carton Style

Corrugated packing materials usually come in five categories: large A, small B, medium C, mini E and extra large K (See chart 1). Their differences can be found in the following chart. The common styles of corrugated cartons have been introduced in the previous section as slotted cartons, telescope cartons, etc, with slotted carton a most common type. Choice of corrugating and carton style can be made according to the information collected in the preliminary stage or by the clients. Micro-corrugated cartons are more suitable for digital products while slotted cartons are the better options for drinks.

Chart 1

Category	Size	Height (mm)	Quantity 300mm
A	large	4.5 - 5	34±2
B	small	2.5 - 3	50±2
C	medium	3.5 - 4	38±2
E	mini	1.1 - 2	96±4
K	extra large	6.6 - 7	24±1.5

Size

Size is a crucial element in the design of corrugated cartons as it includes the internal diameter, outer diameter and working size. The internal diameter (X_1) is decided by the size of the commodity and needs to be able to accommodate the commodity completely. Outer diameter (X_2) is the size taken into account in the circulation process, which affects the space needed in transport and storage. Working size (X) is the manufacturing dimensions during the manufacturing process. The production efficiency of each product can be calculated according to the carton size, working parameter of the production line and die-cutting machine (See figure 10).

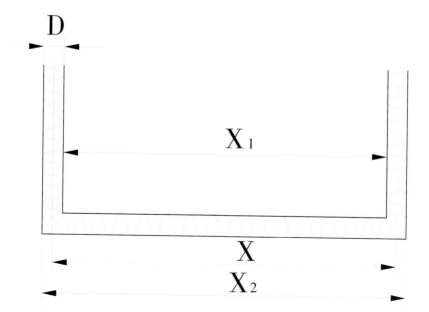

10

Layout

The layout of corrugated cartons including the elements of product names, corporate logos and names, instructions and specification, is decided by the clients. Designers will arrange the information on an aesthetic basis. Colours, shapes, font styling, illustration and crafting can add personality to the corrugated cartons. The finished design needs to be checked carefully to avoid errors that may cause great economic loss in bulk printing. Mass production should only commence after all the design details are checked and confirmed (See figure 11).

11

This is a corrugated carton by jauku design with product name, design and production company, product introduction, and other information on the carton surface.

Package Design for Fragile Items

Fragile items can be categorised as items of fragile nature and items with fragile container. The packaging of these items usually needs gap filling or fixing clamps. Air cushion film, polyurethane foam, air bags and newspaper are often used for this purpose, providing satisfying performance in cushioning, gap filling and wrapping. The corrugated carton on the outside brings strength and durability and can greatly increase the safety level (See figure 12).

Fixing clamps are a form of structural design in which the item is fixed in a certain position to avoid displacement.

Designed by Matt Maurer, this teacup packaging design uses small corrugated cartons matching the size of the teacups and newspaper in wrapping to reduce the vibration impact. It is also important to put the sign of 'Fragile' on the outside of the carton, reminding people to handle with care during transport.

12

Design for Aesthetic Purposes

Corrugated packaging is no longer restricted to the plain yellow box, as designers try to increase the artistic and aesthetic value through graphic design. The production can be carried out according to product features or be processed in a way the designers feel comfortable with (See figure 13-15).

The following are a series of the recycling bins designed by Urska Hocevar. The lovely illustrations on the bins and brightly-coloured covers make them stand out in the office environment. This helps to remind people of the necessity to reduce consumption and waster, increase recycle and reuse, and lead a low-carbon environmental friendly life.

This is the wine package Federico Galvani and Andrea Manzati designed for Christmas. It has the appearance of a book. Considering both aesthetic and technical factors, the designers decided to use corrugated packaging as its material, since it is easy to cut, print, light in weight but rather strong. Though not expensive in cost, it can convey quality and warmth as part of a present.

Colours

Nature provides us a goldmine of colour resources. And colours have a direct and strong impact on people's thoughts and emotions. The stereotype of corrugated cartons no longer exists as designers impose emotions and implications into corrugated packaging via the means of colours and shades (See figure 16-17).

16

This group of tea packaging by Nadia Arioui Salinas employs a series of bright and cheerful colours. On one hand, they stand out from the large variety of commodities and effectively attract the customers; on the other hand, different colours indicate different flavours, helping in the process of visual distinction. Using water colour as the painting colour, the corrugated packaging is biodegradable and can be reused, with impact on the environment minimalised.

This luxurious egg product package by Luz Selenne Guardado is made of corrugated cardboard, protecting the eggs even when they fall. By simulating the pattern of diamond edges, it conveys a sense of class and luxury. The contrasting colours of yellow and grey stand for gold and silver. A bright yellow ribbon is used to seal the package, indicating valuable items inside.

Application of Technology in Corrugated Packaging Design

Corrugated product is strong and easy to mould, it is therefore feasible to add special features by the application of technology (See figure 18-19).

The Michal Kupilik design for the traditiona Czech herbal liqueur employs the hollov technique by carving out the name of th product and the shape of the bottle. I makes a unique package by multi-folding th corrugated cardboard, and when it is opene the bottle will reveal itself. A cocktail recip can be found on each page of the cardboarc so the liqueur can not only be enjoyed on it own but also in a cocktail.

18

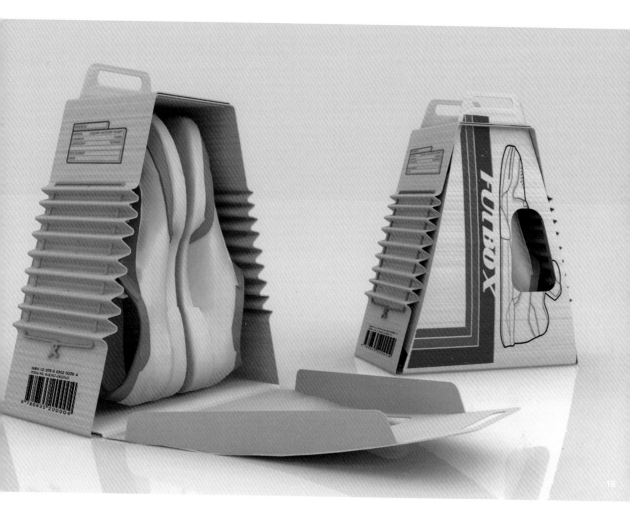

Shoebox is a box in a variety of sizes according to the size of the shoes. This makes the design and production of shoeboxes not very convenient. Turkish designer Sencer Özdemir and Büşra Mehlike Kurt found a wonderful solution using corrugated cardboard. Their idea is to make a shoebox whose size can be adjusted and the key to this design is form a buffer zone on the side of the shoebox using corrugated cardboard folding. It is controlled with a rope, which can be loosened when a bigger box is needed and as it tightens the box is drawn back together to match the shoe. This design makes shoeboxes easier to store, stack and display, and the handle detail on top makes it convenient to carry.

Transport and Carriage

During World War I, corrugated cartons took up only 20% in the transport package, with wooden cases 80%. The percentage witnessed a dramatic increase in World War II to 80% while the use of wooden cases decreased quickly. In less than 20 years, corrugated carton rapidly became a widely used container in transport package, as it is light in weight and durable in use, also convenient for transport. Today, designers try to enhance the advantages of corrugated cardboard and apply them in packaging design (See figure 20-21).

20

This is a football package designed by Ilya Avakov. The three-dimensional design provides a fixed display of the football, and as the package is not sealed off, customers can touch and feel the football before making the purchase. The handle on top is easy for carrying and transport. If you like it, you can carry it home with you. Just that simple!

This serie of wine package designed by Olssøn Barbieri employs illustration of the alluring French elements and a structure convenient for display and stacking. It is easy to carry by hand. Although the commodity has a certain weight itself, the handle helps reduce the stress.

Design for Special Products

For a variety of commodities, safety is a major factor to consider. Corrugated packaging for such items needs to be strong and safe, and at the same time resource-efficient to avoid over-packaging (See figure 22-24).

Take this nail packaging by Pier-Philippe Rioux as an example: to begin with, the designer needs to make sure that people working on the package do not get hurt by the sharp nail tips. Therefore a thick and strong corrugated cardboard is chosen as the packing material; making sure that the nails do not penetrate the outer package. The middle section of the nails is also fixed with corrugated cardboard. By fixing at the tip, middle and tail, the nails are steadily packed in the carton. The heads are used to indicate the model and size of the nails, so people can choose what they need, simply by looking at the outside of the cartons. No glue or ink is used in this package design, and as corrugated products are recyclable and reusable, this design contributes to environmental protection to the highest degree possible.

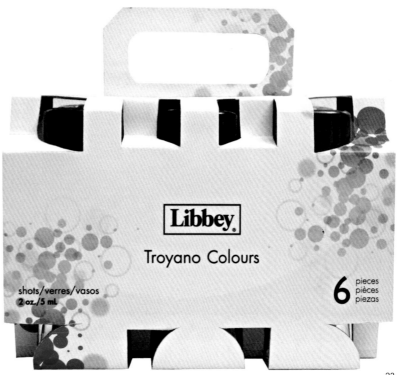

The design of glassware and porcelain calls for more consideration in safety due to their fragile nature. This glass package designed by Li Jing is a one piece corrugated cardboard product. More than 50% of the commodities can be observed from outside the package, which largely increases the item's visibility. The front and back pattern consists of colourful circles, promoting the fun and colourfulness of the small glasses. At the inside of the carton, corrugated cardboard lining helps fix the items and reduce vibration, improving the safety degree. With the handle on top, people can carry a set of small glasses with peace of mind.

Recycling and Reuse

Corrugated cardboard is a renewable resource, which means it can be reproduced, restored and reused in a certain period after the treatment of repeated use, consumption, waste and reprocess. Renewable resources can reduce the exploitation of natural resources, protecting non-renewable resources to the most extent possible, it therefore plays a positive role in energy saving and resource conservation, as well as environmental protection. Designers can also contribute to increasing the efficiency of corrugated products with their designs, so that they not only work as packaging, but in other uses too (See figure 25-28).

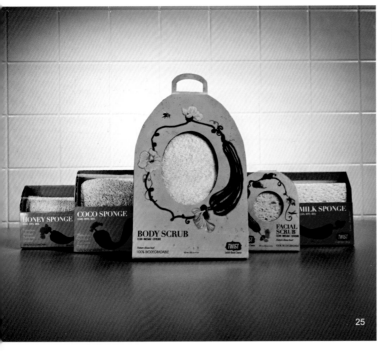

Designed by Boya Liu, this loofah package is semitransparent so the customers can choose the size or shape of the loofah they like. When opened, the package can be separated into a tray for the loofah so it is convenient for use and easy to clean.

25

The Fold Aide Mechanics

1. The ironed shirt is placed face down on the open box as shown.

2. The bottom part of the shirt is folded up till a marked line.

3. The left panel of the box is folded as shown.

4. The right panel is now folded

5. The left panel is then folded once again.

6. Next the bottom panel is folded.

7. The ironed shirt is now folded perfectly.

8. The end result.

This is a work of iron package by Leo Burnett, Mumbai. It resembles the appearance of an iron, and the fonts on the package are pictographic so it is clear that it is an iron product. As for the structural design, the package can serve as a clothes folding board when it is flat. It is a classic example of the multi-use of successful package design.

Printing on Corrugated Cardboard |

Flexography

Flexographic printing applies water-based ink directly on corrugated cardboard, which is also referred to as water-based printing as an additive method of printing. During this process, a negative version of the master copy is produced via an electronic colour scanner and printed via the printer, while the ink is transferred from the anilox to the printing plate, and then to the corrugated paper. Through colour topping and overlapping, the positive image is attained as the reproduction of the master copy and of the original quality.

Since its printing plate is flexible, the pressure of flexographic printing is lighter than that of other printing systems. Therefore it provides satisfying performance in terms of solid ink colour, clean print and higher press runs, etc. In flexography, there is not much specification concerning the glossiness, absorbency, or thickness of corrugated cardboard, therefore it is a common system employed for corrugated product printing. Moreover, water-based ink is easily absorbable, fast drying and eco-friendly, with no toxic volatile gases, especially suitable for food, pharmaceutical and cosmetics packages (See figure 29).

29

Designed by jkr

Offset Lithography

In offset lithography, the image and document are pre-printed, then composited with corrugated material to get coloured corrugated cardboard, which is cut to size to obtain final products of sophisticated pattern. There are multi-colour and multi-functional web and sheetfed presses in the offset lithography category. The web model features high production rate, suitable for pre-printed carton production of relatively fixed structure and large quantity. The top sheet can then be applied to corrugated material. The sheetfed model, however, can work directly on fine corrugated cardboard, providing better accuracy and more stable quality than web offset lithography. It is employed in small-batch carton printing and production with a range of sizes to chose.

Offset lithography produces thin ink coating suitable for products of fine lines and multi-colour net lines. Its advantages include high lattice point resolution, affluent layout, clear boundaries, soft and natural colour, satisfying reproduction degree, easy plate production and low cost. It is mostly used in the production of high quality corrugated packaging for promotional merchandises and gift wrap, etc (See figure 30).

30

Internal design of Nike

Serigraphy

Traditionally, serigraphy is a printing technique using fine woven mesh of silk, synthetic fabric or metal to support an ink-blocking stencil to receive a reproduction of the master. Modern screen printing employs photosensitive material instead. The ink is transferred to the substrate through the mesh under squeegee pressure, forming an exact copy of the original document. Screen printing can be applied directly on corrugated cardboard.

It is a flexible, stable, and cost-efficient printing system. The products are highly competitive as they are of vigorous top-tone, high saturation, and strong visual impact (See figure 31).

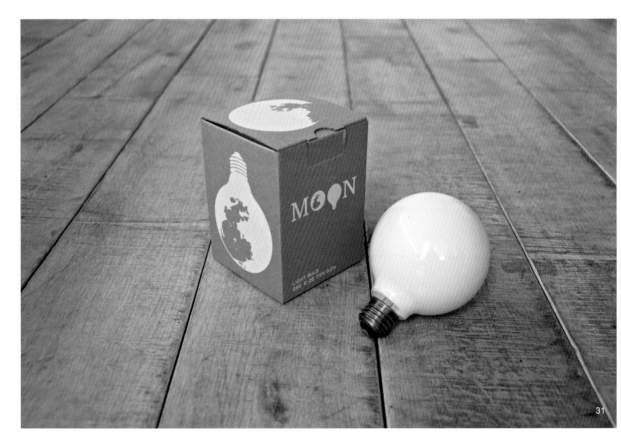

31

Designed by Hidetoshi Kuranari and Chihiro Konno

Rotogravure

There are also web and sheetfed models in a range of colours and specific functions in rotogravure. Producing thick ink coating, rotogravure prints are full and dimensional with a strong sense of layout and texture. Because gravure is capable of transferring more ink to paper, it is a preferable choice for large area or high density printing. In corrugated paper printing, it is a process with the consistent high quality and the least possibility of colour distortion.

Since gravure printing requires the creation of one cylinder for each colour of the final image, it is expensive for shorts runs and best suited for high volume printing (See figure 32).

Designed by Kitchen & Good Wolf

Cutting and Creasing of Corrugated Cardboard |

Die cutting is the main method employed today for corrugated cardboard cutting. Die cutting machine works at high speed and is largely efficient, yet some preliminary work is required before it starts to work:

Design
Tool processing
Creating die
Cutting die fixing
Creasing plate
Cutting and creasing
Bindery

Design

Specific dimensions of the corrugated package including inner diameter, outer diameter and processing diameter need to be confirmed during the design process. Rounded corners are preferable for simpler cutter arrangement, while multiple adjacent narrow edges should be avoided to reduce excess. A test run will be carried out after the details are determined, to check the design from both functional and aesthetic perspectives before mass production begins.

Tool processing

Sharp blades of various hardness and tenacity are used in the die cutting process while crease lines, or creasing rules produce creasing in the material. They are bent into desired shape and mounted to a strong backing before use. The connection of blade and joint, namely the degree of tightness, should be controlled within an appropriate range.

Creating die

After mounted in place, the cutting blade is referred to as a die. It requires meticulous work to create dies so it can cut efficiently with minimal waste. Multiple dies are normally fitted on one mount, to form the desired shape of the final product. This can be done manually, or by machine or laser.

Cutting die fixing

Rubber strips are needed on both sides of the cutting tool to give pressure and keep the blade in place during operation. A base plate is attached to the bottom of the cutting die, leaving only the same depth of all knives revealed to provide a consistent cutting force.

Creasing plate

Then it comes to the setting of creasing plate. The creasing process leaves defined bent marks on the corrugated material, so it is quick and easy to assemble. Depth of the creasing part depends on the compact thickness of corrugated cardboard.

Cutting and creasing

Now the press is ready to roll. Corrugated cardboard is fed into the machine, and the production commences after successful test run is conducted.

Bindery

Development Trend of Corrugated Packaging

Smaller Products and Smaller Batches

According to a designer, if a corrugated carton can be reused five times before disposal, 4,000 trees and 2.4 million gallons of water can be saved, which could produce enough electricity for 49 families to use for a year. The development trend of modern corrugated product industry and technology is directed at high efficiency, multi-functions, environmental protection and resource conservation. Through constant progress and improvement in design and technology, corrugated product maintains its competitiveness and sustainable vitality while demonstrating positive social values. The continuous development of corrugated packaging products corresponds to the need of this environmental friendly age. With the rise of internet shopping, corrugated package production is becoming smaller in batches and more personalised in mode, to adapt to the diversified and personalised internet commodities and custom made products (See figure 33-34).

33

34

This work is designed and produced by Gary Corr, who integrated his resume and some previous works in the corrugated carton and posted it to potential clients. It is more straightforward and convincing than what one can convey on a piece of paper.

Resource Conservation

In this era of green mission and environmental protection, it is a primary task for designers to use the materials on a rational basis and improve the utilising efficiency, in order to achieve the maximum economic and social benefits with the minimal consumption (See figure 35-37).

This skeleton glass package designed by Roy Sherizly is impressively special. It is an environmental friendly and sustainable package especially designed for fragile items. Its major guideline is to get rid of all the external panels in package design and retain the 'skeleton' structure. Even without the external corrugated cardboard, it keeps the integrity of the overall structure and remains protective in shock and vibration. Weight loss improves the transport efficiency and is convenient for stacking.

This bulb package by Oscar Salguero is also a classic in material-saving design. The inspiration comes from the delicate Japanese egg package as the bulb base is fixed in the corrugated structure with extending corrugated cardboard wrapping around the bulb. The structure is slot-fixed without using any glue. The product information can be found on the corrugated package. The bulbs come in groups of three. They can be taken off one by one, without damaging the package for the unused bulbs.

37

Multi-functions

The environmental advantages in corrugated packaging refer to the reusing of the cartons, and moreover other possible functions including clothes folding boards and small pots, so they can continue to function in other uses (See figure 38).

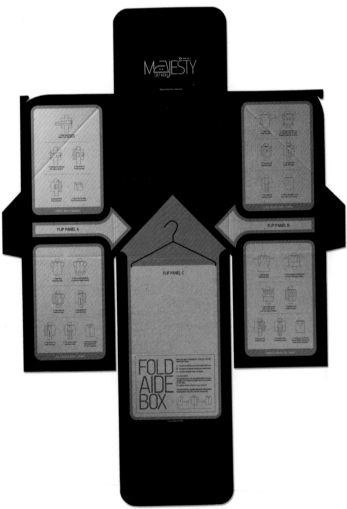

38   *Corrugated carton used as laundry folding board*

Wider Application

Corrugated packaging is a green packaging material, of both value in use and renewable value. With the increasing awareness of environmental protection on a global scale, corrugated products are more likely to be used in all sectors and industries. It is therefore necessary to actively respond to market demand, expand product categories and extend client groups (See figure 39-41).

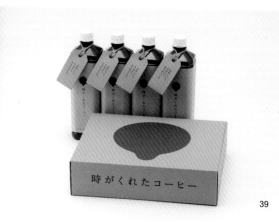

39

40

41

Corrugated cardboard is widely used in all sectors of daily life.

Fermented Butter Cake

This is package of simple tasty butter cake using fermentation butter which is unusual in Japan. The cakes are baked in a farm facing the Okhltsk Sea. COMMUNE represented the simple mild tasty cake using corrugated cardboard box wrapped with cake illustrated craft paper.

Design Agency : COMMUNE
Production Date: 2010
Client : North Plain Farm
Nationality: Japan

BBQ Sauces

Different people have different tastes, so the designer decided to create smaller packages of sauces that would allow bringing sauces with different tastes with you on a camping instead of one big bottle of a BBQ sauce. The keywords for design: Raw, American, Vintage. American and vintage feeling the designer tried to achieve by using specific fonts. Raw feeling she tried to achieve by using rough texture of corrugated paper. The design didn't use pictures, it was important that 'BBQ sauces' would catch an eye. The use of the rough papers and cardboard gives the necessary associations with the camping.

Designer: Olesya Kurulyuk
Production Date: 2011
Nationality: Ukraine

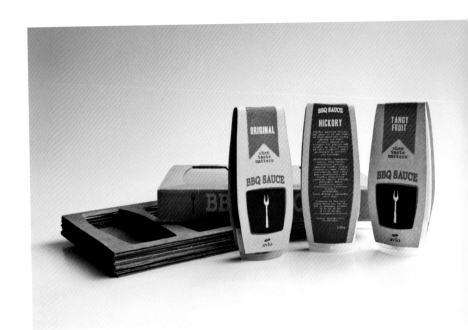

Rubén Álvarez
Aula Chocovic
Fòrum Gastronòmic
Girona'09

Nº 120/120

Code Egg

Packaging design for the limited edition chocolate work of art, 'Huevo Código'(Code Egg) by the chocolate artist Rubén Álvarez. The corrugated package conveys a rustic feeling.

Design Agency: Zoo Studio
Production Date:2009
Client: Rubén Álvarez | Aula Chocovic
Nationality: Spain

Happy Sweet New Year

You don't have to be rich to be a lord. Under this premise the small workshop 'Senyor Estudi' –Lord Studio– is born. It is an enterprise that defends clear values, opting for quality and made-to-measure projects over quantity and standardization. Their identity is formed by an assortment of moustaches printed on coupon book paper, a paper that is fast disappearing but which was greatly used in the last century. All the stationery and packaging are obtained using a single printing plate and different coloured papers and corrugated material. Senyor Estudi plays with different types of moustache design to make a Christmas souvenir that makes you want to lick your moustache.

Designer: Lluís Serra & Mireia Sais
Production date: 2011
Client: Senyor Estudi
Nationality: Spain

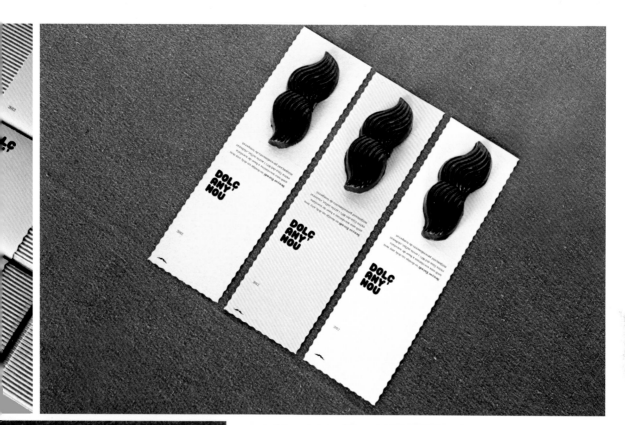

Szelet

The designer wanted to choose a strong concept which can be developed into a franchise later. Therefore the designer has chosen a very intense colour – red – and a familiar shape that evoke the 'world' of pizzeria and fast food restaurants. Corrugated paper is used as the packaing material.

Design Agency: kissmiklos
Production Date: 2011
Client: Szelet Pizzeria
Nationality: Hungary

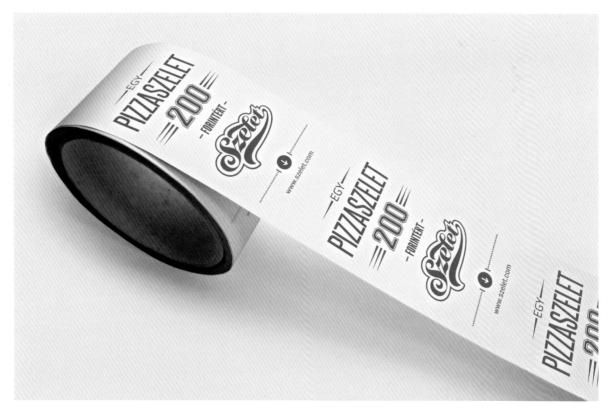

Pizza Nova

The new identity is not only visually unique amongst its competition, it speaks to the distinctive quality and authenticity that this product delivers within its sector. The corrugated packaing can block off oil effectively.

Designer: Boya Liu
Production Date: 2006
Client: Pizza Nova
Nationality: Canada

Egg Box

Material used to produce this product is natural micro-corrugated carton and as a complementary material rubber. The eggs placed into an ellipsoid, cut from the carton. The egg holder is open from the top, but if its turned upside down, it will still hold firmly the eggs, they will not fall because of the elasticity of the rubber.

Designer: Éva Valicsek
Production Date: 2010
Nationality: Hungary

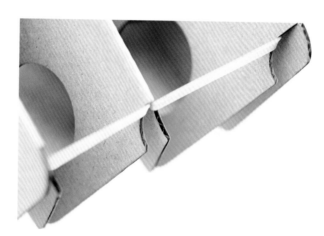

Egg Box 1

Micro corrugated cardboard package for 6 normal eggs. The package is from one piece, space-saving, without glue. The closing is solved by two snapping handles which makes the package and the eggs stable. You can open the box by ripping the perforation.

Designer: Anita Vaskó
Production Date: 2011
Nationality: Hungary

Egg Box 2

The package is from two pieces of corrugated paper, space-saving, without glue. One part from cells and the other one is for closing. The cells are use to hold the eggs and the closing part stabilizing them. You can take out the eggs by pressing the snapping edge on the bottom and the eggs will emerge from the package.

Designer: Anita Vaskó
Production Date: 2011
Nationality: Hungary

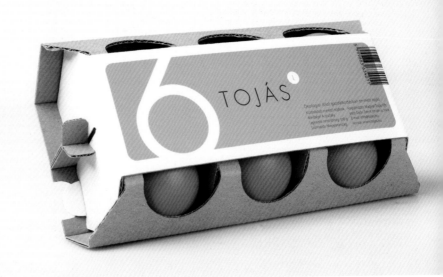

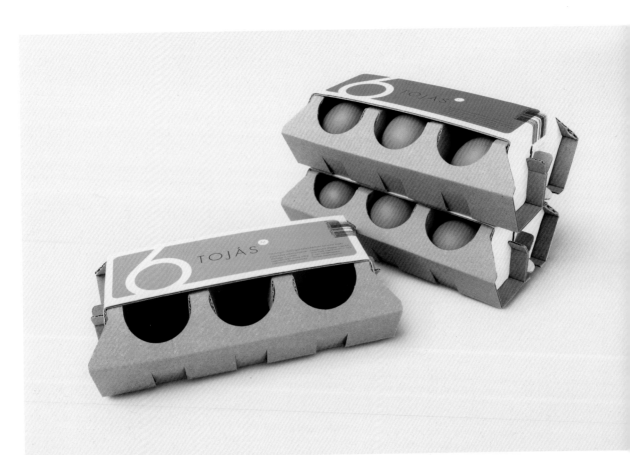

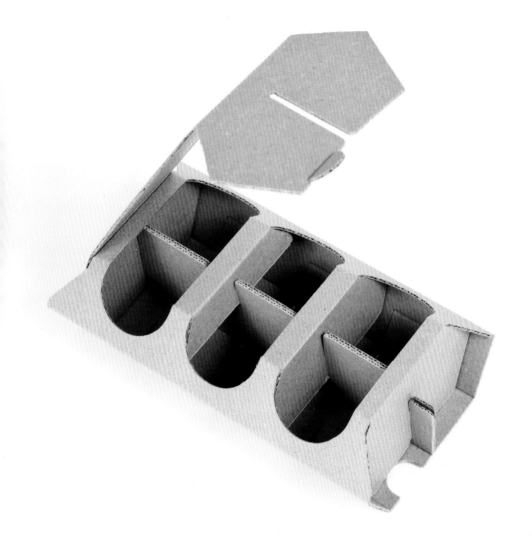

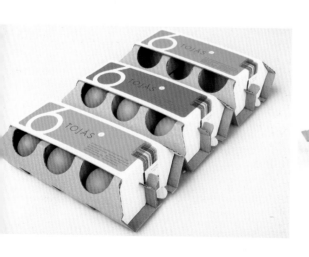

A Single Egg

The concept of the corrugated packaging is to replicate the action of laying an egg, this action brings the fun back into obtaining the egg from its packaging.

Designer: Tim Sumner
Production Date: 2011
Nationality: UK

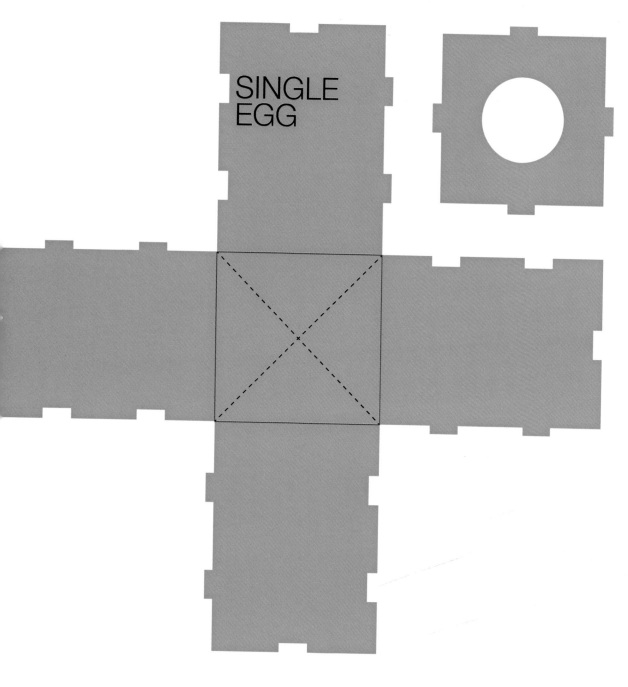

SINGLE
EGG

'Apocalypse:
Emergency Boxes'

A collection of mass produced
corrugated fiberboard boxes
intended to be filled with flyers
and emergency information in
times of catastrophe. The project
injects humour into information
and facts that are usually used
by the popular media as a tool to
invoke hysteria.

Design Agency: GLD/FRD
Production Date: 2011
Nationality: Israel

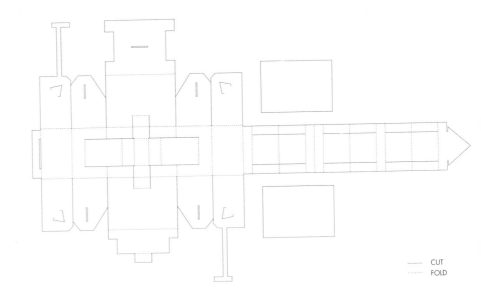

—— CUT
········· FOLD

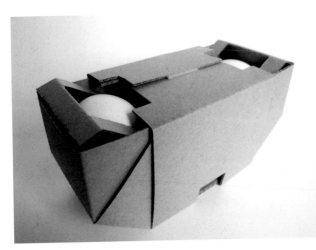

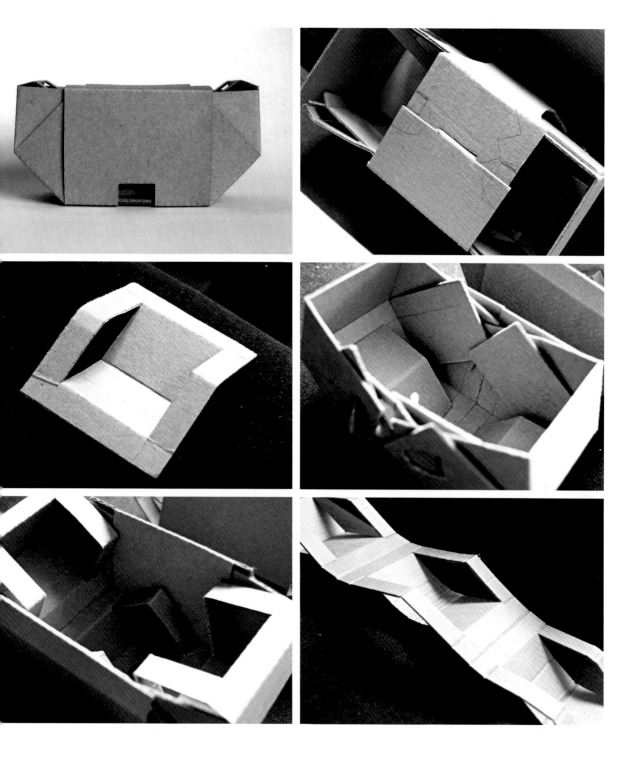

6 Brown Eggs

Brand and Sustainable Package Design concept for 6 extra large, organic brown eggs. The package is envisioned to be printed with soy based inks on 100% post consumer corrugated paper board, an environmentally conscience alternative to a plastic incased egg package.

Designer: Sarah Machicado
Production Date: 2010
Nationality: USA

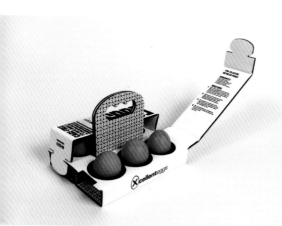

Nestbox

The inspiration for the design came from the nests of birds which are the original place of eggs. The basic decision was to construct an egg box without any sticking; only jointing. Also the goal was to leave the eggs visible and also safe. After all, the whole box was cut out from one piece of corrugated cardboard by laser. The results become a functional package which refers to the solutions of nature.

Designer: Dénes Janoch
Production Date: 2011
Nationality: Hungary

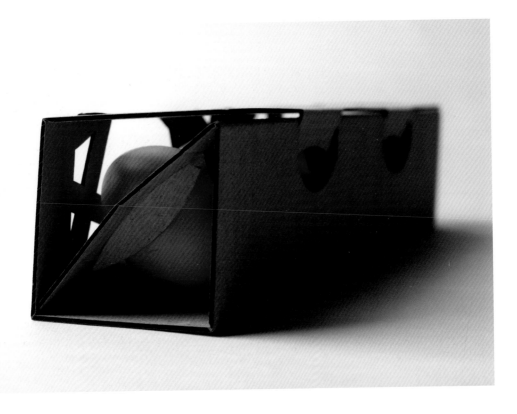

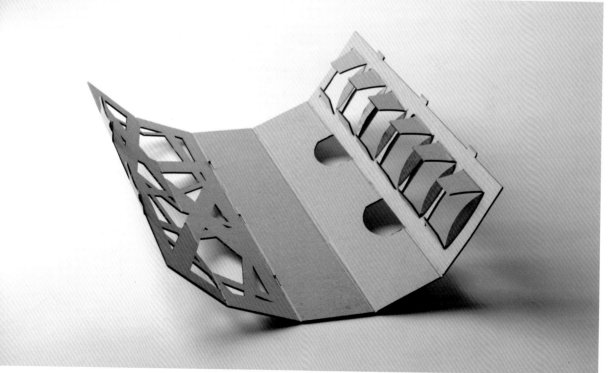

Vingulmark

The designer chose an exclusive
Scandinavian design statement
to emphasize and communicate
the clear and cold Nordic
climate that the hens live in. The
recyclable corrugated package
contains lots of information to
provide customers who are
concerned with the purity and
organic food, a confirmation of
the choice they make. Farms that
are producing organic eggs are
carefully selected. On each egg
carton, you can read and see
pictures of the farmer, chickens,
which area the eggs comes from.
This emphasizes how close the
product is from the environment.

Design Agency: Ghost
Production Date: 2010
Client: Vingulmark
Nationality: Norway

My Lovely Tea Time

The design of this product line, which consists of two teas in special packaging and two thermoses, was conceived in a floral style. Specific lettering crowns the packaging's graphics. Corrugated package provides the products with protection during transport.

Design Agency: o zone
Production Date: 2011
Client: Neavita - HP Italia Srl
Nationality: Italy

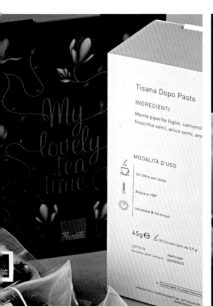

Southern Comfort Fiery Pepper Launch Kit

Cue created a launch kit to promote the product to the industry trade. The corrugated box contains branded coasters and a siren to deliver a disruptive, high- energy teaser. The launch kit established a visual language and system of iconography that was embraced by agency partners and used to promote Fiery Pepper to consumers and bartenders.

Design Agency: Cue
Production Date: 2011
Client: Brown-Forman
Nationality: USA

Gustas

Packaging design of imported wholesale beef. The challenge was to create an original image with great visual impact, especially with piled-up boxes, as it is addressed to wholesale market. The result shows originality, impact and exclusivity. The application of corrugated material is cost saving.

Design Agency: 2creativo
Production Date: 2007
Client: Sucarn
Nationality: Spain

Teapot Packaging

The Teapot products consist in a gourmet and organic tea line, which is conscious with the environment. The materials used are low environmental impact. The corrugated packages have been created to distribute in specialty shops and boutiques. Another type of package is also created to store the product in restaurants. This is the designer's final project at EASD Valencia (school of art and design in Spain). Following the basic rule of something ecological and natural Nadia Arioui Salinas chose hand-painted illustration. Even the paint (watercolour) and the packages were biodegradable and reusable. The designer tried to minimize technological intervention to create a handmade product.

Design Agency: Nadia Arioui Salinas
Production Date: 2010
Nationality: Spain

Tea Package – Chinese Tea Traditions

The recyclable tea package is made from two single pieces of corrugated paper bended and compounded together without any glue and one printed etiquette with various Chinese traditions such as tea for family gatherings, tea to honour someone or tea to heartful excuse. This package was made for the Young package competition.

Designer: Michal Marko
Production Date: 2012
Client: Young Package
(Competition)
Nationality: Slovak

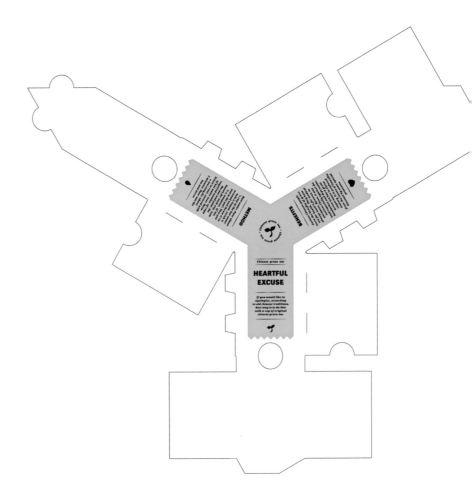

E·Oliva

Both corrugated cardboard and linen thread provide the pack with a natural spirit, taking care of the small details in such elements like the handcrafted sewing or the soft relief in the typography. At the same time, the message is shown in a honest, simple and forceful way in every aspect, with a typographic composition that contributes to show a contemporary look to the pack. The family tradition related with E·oliva provides great respect to the environment. That is why the pack does not use any type of adhesive and all the materials are biodegradables.

Design Agency: Alberto Aranda Design
Production Date: 2010
Client: E·Oliva
Nationality: Spain

SIZ Box

A renowned Italian printing
company desires to send its
client two bottles of wine as a
present for Christmas time. The
designer decided to customize
the package to match some
relevant issue: book-like visual
aspect; use of corrugated
cardboard. It can be easily cut,
printed, glued; it's light but
strong, to be easily shipped; it's
not expensive and gives a warm
feeling when in hands.

Designer: Federico Galvani,
Andrea Manzati
Production Date: 2011
Client: SIZ
Nationality: Italy

Prole Piss Brewery
Split Pack Beer

This packaging represents the author Raymond Carver's working class attitude and writing style, as well as his struggles with alcoholism. This packaging was constructed out of one piece of di-cut and screen printed corrugated cardboard that can hold around 6 bottles of beer when folded.

Designer: Sarah Sabo
Production Date: 2011
Nationality: USA

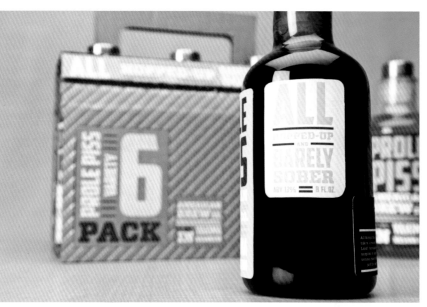

Toki Ga Kureta Coffee

This product is a corrugated
package of four coffee bottles.
The box is adorned only with
the logo and product name for a
simple and sophisticated effect.

Design Agency: COMMUNE
Production Date: 2009
Client : infini coffee
Nationality: Japan

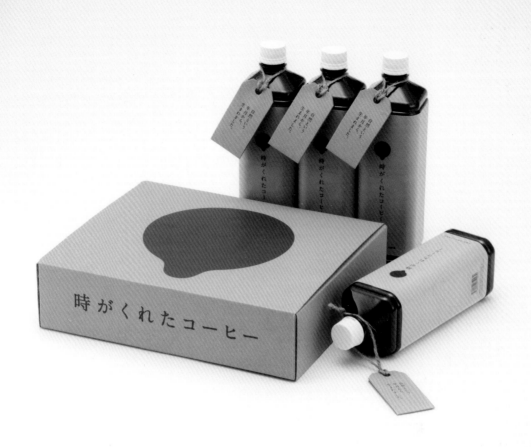

時 が くれた コーヒー

Calling Card

Descriptors for wines often mirror the way we talk about women's tantalising qualities. The choice of corrugated cardboard helps guarantee product safety during transport. By combining the calling card concept with corrugated package, the designer succeeds in making the collection eye-catching and all the more intriguing.

Design Agency: Boldinc Creative Brand Consultants
Production Date: 2010
Client: Saint and Sinner
Nationality: Australia

RelajarTE (relax tea)

Illustration and packaging design for a Tea box. The box was a gift for doctors and health professionals. estudioVACA illustrated the complete team of professionals who work at the process of elaborating medicaments.

Design Agency: estudioVACA
Production Date: 2011
Client: Cinfa laboratory
Nationality: Spain

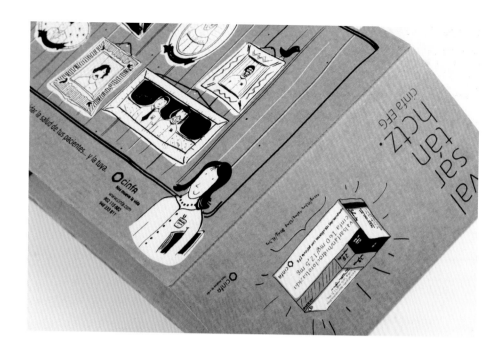

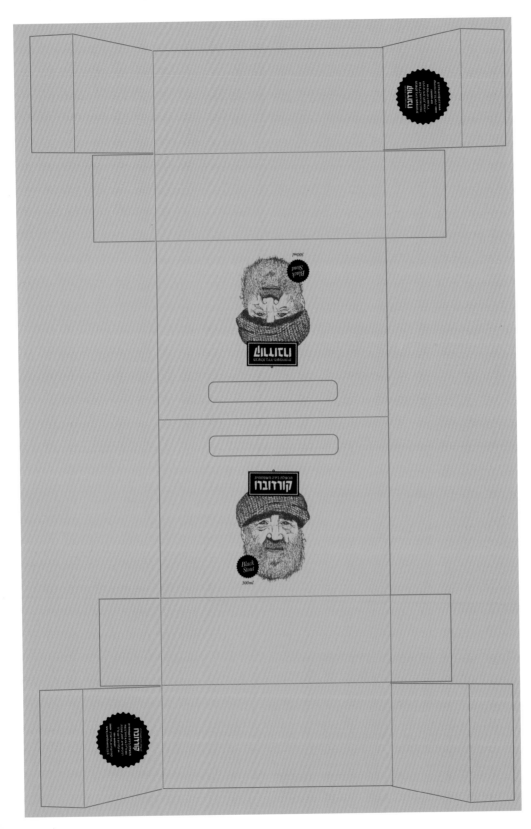

Cordovero Brewery

CORDOVERO BREWERY is an Israeli family beer boutique. Established at 2010. The first brewery factory was located on a street in south Tel Aviv named CORDOVERO. (Moses ben Jacob Cordovero was an important jewish figure in the historical development of the Kabbalah). The brand and design is based on a rough, yet friendly brand character, a nameless old man, printed on corrugated package.

Design Agency: EB Eyal Baumert Branding studio
Production Date: 2010
Client: CORDOVERO BREWERY
Nationality: Israel

Golan Brewery – Bazelet

Branding and design for the Golan Brewery. The Golan Brewery was launched as a joint venture by the Ohayon family and the Golan Heights Winery. Located in Katzrin, the brewery produces beer of an international caliber. In the search for a design concept, the designer engaged in a dialogue between the local and the sophisticated, the masculine and the liberated, and the rugged and the cultured. The package is made from corrugated cardboard, for it is easy to cut and craft.

Design Agency: Blend-It Design
Production Date: 2010
Client: Golan Brewery
Nationality: Israel

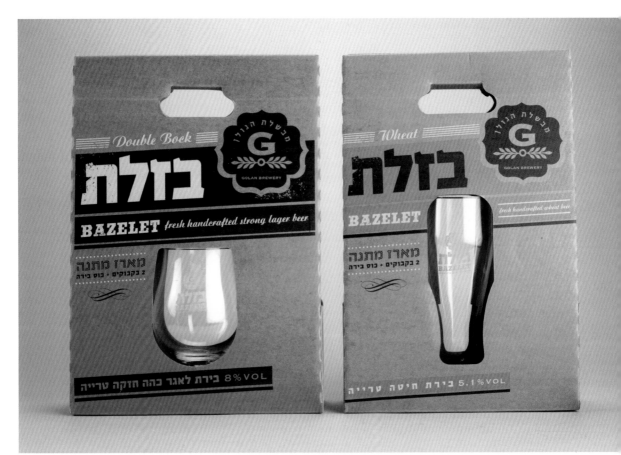

Wildflower Whisky

An innovative, new brand of
Scotch whisky that challenges
the conventional masculine
imagery surrounding the drink,
appealing to a female consumer.
'Wildflower' is a premium
blend of single malt whiskies,
sourced from small, independent
Highland and Lowland
distilleries, and made using
the finest natural ingredients.
The packaging and branding
features an intricate floral pattern
handrawn by Indra Waughray.
The product is inspired by nature
with packaging designed to have
a reduced impact on
the environment. The box is
made from laser cut recycled
corrugated cardboard and
the labels laser etched on
handmade, recycled, plantable
paper containing live wildflower
seeds.

Designer: Indra Waughray
Production Date: 2012
Nationality: UK

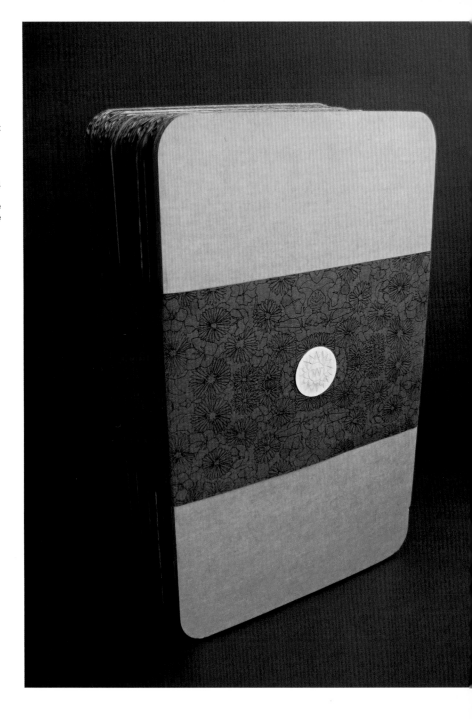

This packaging is made from recycled and sustainable materials and we encourage you to recycle it again. You can also peel the labels from the bottle and re-use it yourself. The labels are laser etched on paper containing live wildflower seeds. Soak them in water overnight and plant them to grow some beautiful flowers and give a little something back to mother nature!

La Vallée des Rois

Input
A new BiB with white wine from La Vallée des Rois, in the Loire Valley, France. 'The Kings´ Valley' is well known for its historical architecture, ancient towns and beautiful castles from the 'l'Ancient Régime'.
Output
 The designers signal traditional French royal values with a grand illustration of a castle in an almost 3-dimensional design solution. The use of corrugated material provides safety and protection during transport.

Design Agency: Neumeister Strategic Design AB
Production Date: 2010
Client: Berntson Vin
Nationality: Sweden

Orrefors - You/Me

Input
Orrefors is associated around
the world with unique glass
articles, art glass, and custom
glass. A classic Swedish design
company. This time, however,
the designer was a student of
University College of Arts, Crafts
and Design in Stockholm and
with his series You/We he wanted
to attach a younger target group.
Output
The Design solution emphasizes
the idea of us being individuals,
but at the same time always part
of a family, group... crowd. The
designer choosing pins as the
classic young symbol of sharing
an opinion or liking. To quote
Cassius Clay: 'Me, you... we.' In
general a modern design solution
for a glass that is supposed to
have a life outside of the cabinet.
Outcome eaunched in stores all
over the world and received a
great deal of attention. The use
of corrugated material provides
safety and protection during
transport.

Design Agency: Neumeister
Strategic Design AB
Production Date: 2010
Client: Orrefors
Nationality: Sweden

Becherovka Original
Gift Pack

Package for the traditional Czech liqueur Becherovka was cut out from one piece of corrugated cardboard and additionally serves as a cocktail recipe list. It was presented at the Young Package competition in 2011.

Design Agency: FMK UTB
Production Date: 2010
Nationality: Czech Republic

Promenade

Olssøn Barbieri distilled and strengthened the Promenade concept with a selection of charming and positive French clichés; A walk through France. Souvenir de Paris. Romance, elegance and charm. Most of the text on the corrugated gift box, is handwritten to strengthen the romantic and authentic expression.

Design Agency: Olssøn Barbieri
Production Date: 2012
Client: Vinordia AS
Nationality: Norway

Factor 5.5 Package

Concept package created for the mexican drink made with tequila Factor 5.5. Made of corrugated cardboard and laser cut, in just one piece and without colour applications to highlight the colour of the beverage and the brillante of the glass bottle creating a contrast between the glass and the cardboard. The intention of this package is to be used in stores for display and protector of the product. Due to the form of the package, it can be placed in a corridor and be seen from two different angles to create a game between the observer and the product. The package is designed to highlight the product from all the beverages on a store respecting the simple design style of Factor 5.5.

Designer: Luz Selenne Guardado
Production Date: 2010
Nationality: México

In Vino Veritas

Ink, water, brush miracle begins
which you can only observe and
admire . Wine production is a
unique miracle which can be
experienced by everyone. The
comfortable corrugated package
is an invitation to try and a story
of the future experience.

Designer: Gaze Olga
Production Date: 2011
Nationality: Ukraine

Eco Labels

'Eco Labels' is a student's work executed for the packaging studio in Academy of Fine Arts in Poznan. Joanna Angulska has designed labels of three selected eco products (milk, Himalayan nuts and spelt wheat cookies) for one kind of consumer's group. Typographic labels feature the origin of every single product e.g. village, name of cow or bakery's address. The package of corrugated material conforms to the design concept of nature and ecology.

Designer: Joanna Angulska
Production Date: 2012
Nationality: Poland

Two Hoots

Each owl character has been designed to match the characteristics of the wine, tying into the tongue in cheek descriptions on the back of each varietal. Cabernet Rose: 'Like the screeching owl, this cabernet rose has a rather flirtatious character. Fresh peach and pear aromas combine with cheeky cherry flavours to add sweetness to the playful, fruity wine.'

Design Agency: Maegan Brown
Production Date: 2012
Nationality: Australia

Louie

In Norway hunting is a popular sport. Olssøn Barbieri were asked to design Bag in Box for wine to meet the request for a well paired wine to go with the seasons game. Olssøn Barbieri got inspired by old ads for hunting equipment from the start of the 19th century and combined this layout with ink drawings, hunting ethics, sayings and various information related to hunting. Louie is named after a notorious local Gordon Setter gun dog and corrugated cardboard is chosen as the packaging material.

Design Agency: Olssøn Barbieri
Production Date: 2010
Client: Arcus Wine Brands
Nationality: Norway

Collection Ripasso

The original artwork is hand drawn (ink on paper) and evokes associations to a traditional and opulent Italy. The red ribbon was simplified and tuned to a more vivid red while Olssøn Barbier introduced a golden brown hot-foil in tone with the illustration to achieve a discreet and exclusive expression. Corrugated cardboard of strength and durability is used to bear the weight of the wine.

Design Agency: Olssøn Barbieri
Production Date: 2012
Client: Symposium Wines
Nationality: Norway

My World

Olssøn Barbieri decided to wrap each product in its own origin accompanied by a selection of illustrations of local flora/fauna and culture. Inspired from old illustrated and coloured maps from the 16th century, Olssøn Barbieri created a series of hand drawn maps coloured by aquarelle that embraces each corrugated box. Olssøn Barbieri wanted to create a modern packaging without choosing a minimalist direction but with a completely handmade artwork.

Design Agency: Olssøn Barbieri
Production Date: 2012
Client: Arcus Wine Brands
Nationality: Norway

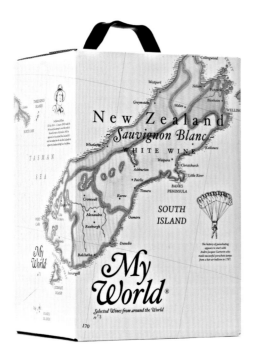

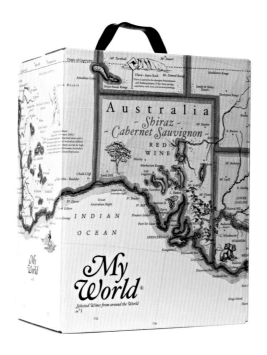

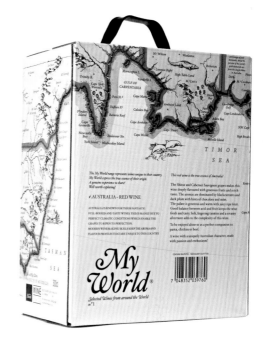

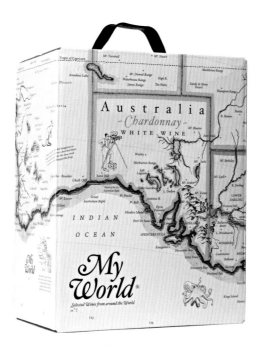

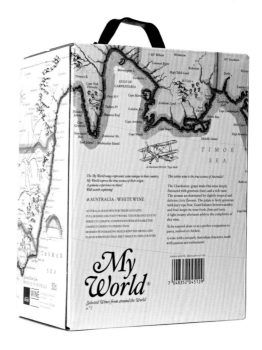

Lewis Vodka

For this school project, students were free to design the packaging of a product of their choice. The only restriction was to make a gift set. Élyse Levasseur decided to design the packaging of a fair trade vodka bottle and its shooters. It was a product that she had never worked with. As organic and fair trade respect nature, it was important to make an eco-friendly packaging, in order to be coherent with the nature of the project.

To do so, Élyse Levasseur used recycled corrugated cardboard boxes in which she made the same die-cut. Once the panels of the boxes are stacked together, it becomes easier to put the bottle and shooters inside; an easy way to wrap something up! In a consumer society where the 'throw-away' is excessively present, as designers, we have the obligation to think intelligently in order to reduce all this waste. That's the designer's way of thinking and this packaging demonstrates that as well.

Designer: Élyse Levasseur
Production Date: 2009
Nationality: Canada

IZZE Holiday Crafting Challenge Kit

IZZE and Olson partnered with some amazing craft bloggers to put on the IZZE Holiday Crafting Challenge. Ten chosen craftsters were presented with this corrugated box of goodies with which to make their IZZE-inspired holiday craft. The contest was a hit on the craft blogs and was a joy to be a part of.

Design Agency: Olson
Production Date: 2011
Client: IZZE Sparkling Juice
Nationality: USA

59th & Mad

59th & Mad, a series of flavoured beers, was created in celebration of all things rock 'n' roll. The designer drew inspiration for this project from a teacher at the Creative Circus, Mr. Dan Balser. As an avid fan of 90's rock, a lover of New York City and the advertising days of old, Dan's personal style provided the perfect starting point to create a beverage meant to be shared with loved ones, and to evoke the true spirit of rock. The corrugated packaging material conforms to the nostalgia theme as well.

Designer: Amy Frischhertz
Production Date: 2011
Nationality: USA

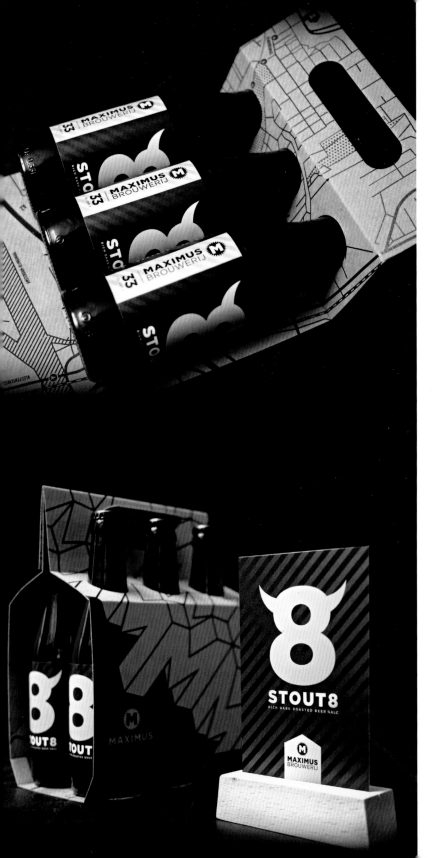

Beer Carrier, Stout 8 label, Maximus Brouwerij

The inside and the outside of the corrugated carrier are screen printed. The inside map shows the biking lane route from the Dom Tower at the heart of Utrecht, to the Brewery that is located outside the city borders.

Design Agency: Leffe Goldstein Graphics
Production Date: 2012
Client: Maximus Brouwerij, the Netherlands
Nationality: The Netherlands

Busch Beer Packaging

'Busch Beer: A classic taste for the modern man.' For this project, the designer was required to find a product and design a new way to package it using modern, relevant design. The designer chose Busch Beer because of its status as a poor quality beer, its popularity among beer drinkers and the fact that its current packaging has a dated look. The designer modernized the brand and created a new way to box the beer, choosing a four pack instead of the typical 6, 12, or 18. The end product resulted in a modern, visually stimulating brand, and a clever way to package the now high-end, craft beer in corrugated material.

Designer: Jacob Weaver
Production Date: 2010
Nationality: USA

Black & Wine

Simple and aesthetic wine bottle box concept made from corrugated two colour cardboard material. Basic squared layout results with an unusual, eye-catching form where packaging structure meets architectural structure.

Design Agency: Istragrafika Company
Production Date: 2010
Nationality: Croatia

Doss Blockos

To capture the brand essence through physical packaging, logo and artwork that was derived from the story and product concept of Doss Blockos. It needed to reflect the art, music and culture, take inspiration from the arts culture imbedded in the underground squat community of New York during the 1990's.Each corrugated box contains half a dozen beers.

Design Agency: Big Dog Creative
Production Date: 2010
Client: East 9th Brewing
Nationality: Australia

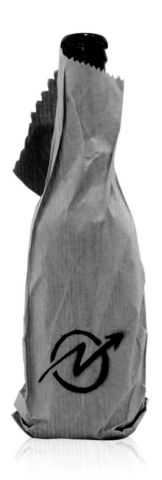

JAQK Cellars

Hatch Design did all design work and sourced all materials, going as far as finding a specialty glass boutique in Milan to create the customized bottle for the flagship Cabernet. To round out the experience, Hatch Design designed a series of corrugated gift boxes, apparel and a limited edition deck of JAQK playing cards.

Design Agency: Hatch Design
Production Date: 2007
Client: JAQK Cellars
Nationality: USA

BH

Designed by BH, the wines of Bodegas Barahonda, with a specific premise: it must position both brands: the product and the bodega. The letters BH stand out in a modern composition in which elegance, simplicity and balance predominate. The choice of corrugated cardboard guarantees product safety during transport.

Design Agency: BOLD BUREAU – SPAIN
Production Date: 2010
Client: BARAHONDA
Nationality: Spain

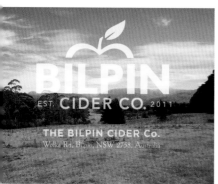

Bilpin Cider Co.

The challenge was to draw on the beautifully honest story behind Bilpin Cider – hand-picked & freshly crushed apples grown in the orchards that surround the township. The simplicity of the process called for a design to match. A vibrant 'Granny smith green' colour scheme combined with a quaint, farmer's market layout, Bilpin Cider represents the new breed of Australian cider. The corrugated package is durable and cost-efficient.

Design Agency: Boldinc Creative
Brand Consultants
Production Date: 2012
Client: Bilpin Cider Co.
Nationality: Australia

La leche 2012

This is a corrugated package design for a Christmas present. It was important to highlight the contrast between wine and milk, white and black.

Design Agency: estudioVACA
Production Date: 2011
Client: estudioVACA´s clients and friends
Nationality: Spain

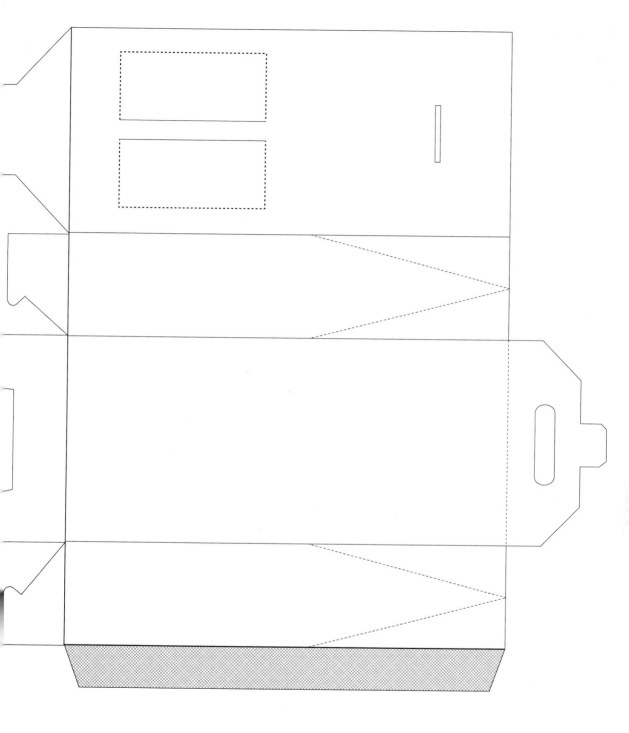

DAB Beer

DAB 3 beers package is a micro corrugated cardboard package for three 0.33 l beers. The package is from one piece, space-saving, without glue. The circle and elliptical shaped holes keeps the bottles in one place, the product won't get damage. The package is made to satisfy both the suppliers and the customers' requirements. The reproduction makes only a tiny amount of waste so it's eco-friendly and cheap.

DAB 6 beers package is micro corrugated cardboard package for six 0.33 l beers. The package is from two pieces, glued on the bottom and the layout is simple. The circles and the handle which separates the two parts keeps the bottles in one place, the product won't get damage.

Designer: Anita Vaskó
Production Date: 2011
Nationality: Hungary

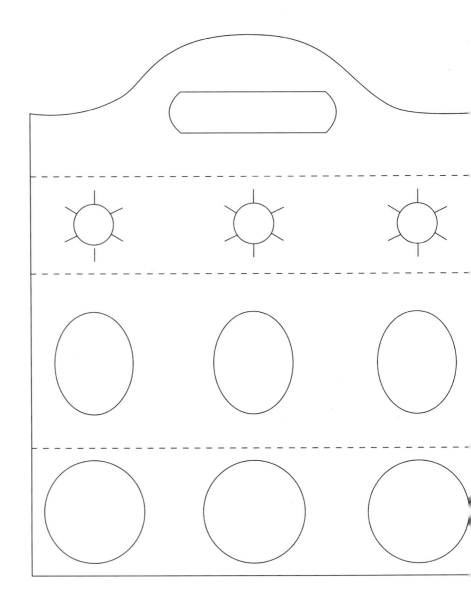

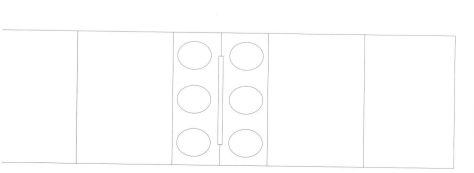

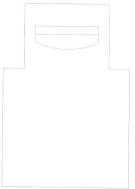

Fragile Collection

The product comes in one piece and by breaking it, you make it useful. The packagings are very simple (yet noble) and communicate to emphasis this unique point. Corrugated cardboard is used to provide protection during transport.

Design Agency: Say – Brand strategy & expression
Production Date: 2011
Client: Studio Kahn – industrial design
Nationality: Israel

This striking ceramic heart pendant, hiding love's secrets cunningly inside it, is yours, first and foremost, to break in two. In a unique fusion of co-creation with the welcome release of your romantic instincts, we invite you to play your own part in the Fragile Collection. Let the heart be struck as by a thunderbolt, discover the power of magnetic attraction.

1. **Hold** the pendant with both hands (keeping your fingers clear of the breaking area).
2. **Snap** in two with a folding action. It will naturally break properly.
3. **Extract** the hidden chain, pulling it until it extends about 45cm (18").
4. **Wear** it around your neck and let the magnets connect together.

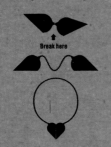

Break here

StudioKahn
contemporary design house

BREAK
A HEART
TO MAKE IT
WHOLE

The FRAGILE Collection

HEART NECKLACE

HEART NECKLACE
The FRAGILE Collection

☐ White / Silvery chain / Gray felt
☐ White / Golden chain / Red felt
☐

Random Series

This piece of self-promotion continues to deepen the concept of random that applied to the corporate identity. Dosdecadatres designed a series of graphic elements and then by a random process, they generated a T-shirt and three posters. The choice of corrugated packaging material adds to the casualness and amusement of the design.

Designer Agency: Dosdecadatres
Production Date: 2009
Client: Dosdecadatres
Nationality: Spain

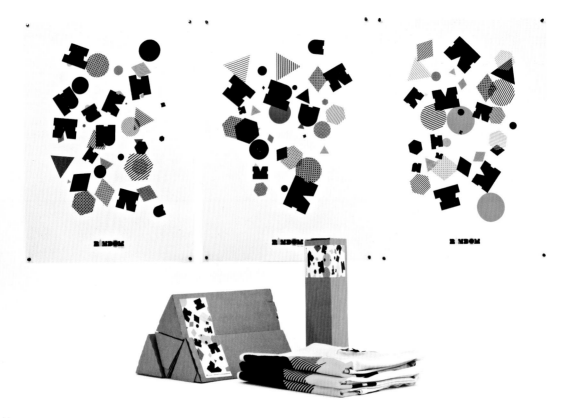

AJVII Year of the Rabbit

It seemed rather fitting that the second release of the 2011 Year of Rabbit collection tie back to MJ's long established relationship with Bugs Bunny. Accordingly, it was decided to revisit the original Hare Jordan colourway, but with some subtle China influenced updates as well as a few carry over details from the AJ2011 release just prior. The package is made from corrugated paper of durable texture and good printability.

Design Agency: Internal
Production Date: 2011
Client: Nike
Nationality: USA

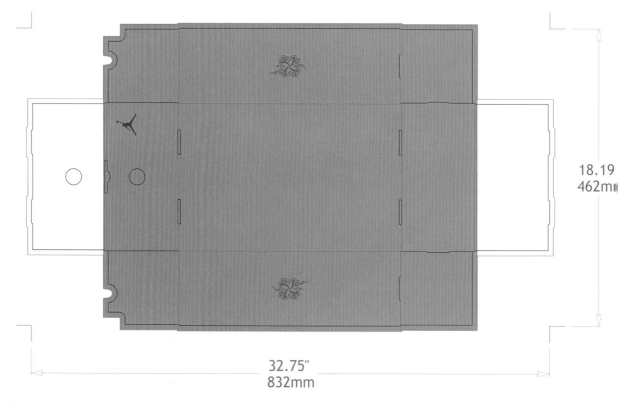

18.19
462mm

32.75"
832mm

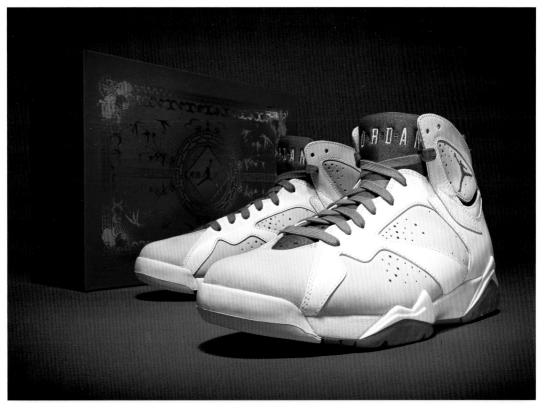

T-shirt Box

The box was designed as a packaging solution for the Leffe Goldstein T-shirts. Each box contains a shirt and booklet that tells a story about the character printed on the shirt. The box is made of 3 mm corrugated board. The top and bottom of the box were separately printed in just one colour, to give it its distinctive look.

Design Agency: Leffe Goldstein
Production Date: 2011
Nationality: The Netherlands

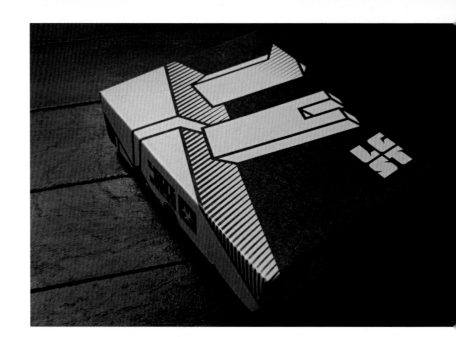

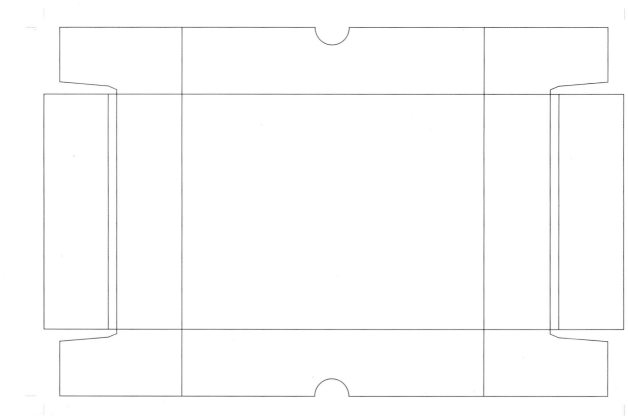

Collapsible Shoebox

This concept shoebox is collapsible and can be returned from its end user to its original manufacturer or distributer for further use. Companies that adopt this system of corrugated packaging will contribute less waste, save money, and become more sustainable. With the right materials and design, this system of packaging can be implemented for various kinds of packaging.

Design Agency: Concept Piece
Production Date: 2011
Nationality: USA

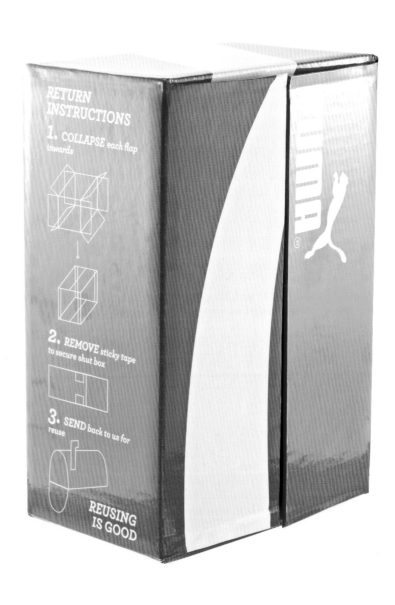

RETURN
INSTRUCTIONS

1. COLLAPSE each flap
inwards

2. REMOVE sticky tape
to secure shut box

3. SEND back to us for
reuse

REUSING
IS GOOD

The World is Watching

HORT designed the type treatment, a font, several shoeboxes (for each part of nyc) and a huge mural as an announcement for 'The world is watching' street basketball event. This corrugated package design was developed in collaboration with Michael Spoljaric from Nike.

Design Agency: HORT
Production Date: 2010
Client: Nike
Nationality: Germany

ignore Moscow

ignore Moscow limited edition is package for independent art project of 5 T-shirts with different Artists' prints. Craft corrugated cardboard package with one colour stencil print including pattern and logo. Developed exclusively for ignore project.

Designer: Konstantin Kolyubin, Lasha Kasoev
Production Date: 2007
Nationality: Russia

Vans

The new box design uses less corrugated cardboard than regular shoeboxes, and is size flexible, in that by varying the way the shoes overlap. In addition to being more efficient, the new box design is more appealing visually, both in that it is more dynamic in shape than rectangular boxes, and displays graphics more appealingly.

Design Agency: Nate Eul
Production Date: 2011
Client: Vans
Nationality: USA

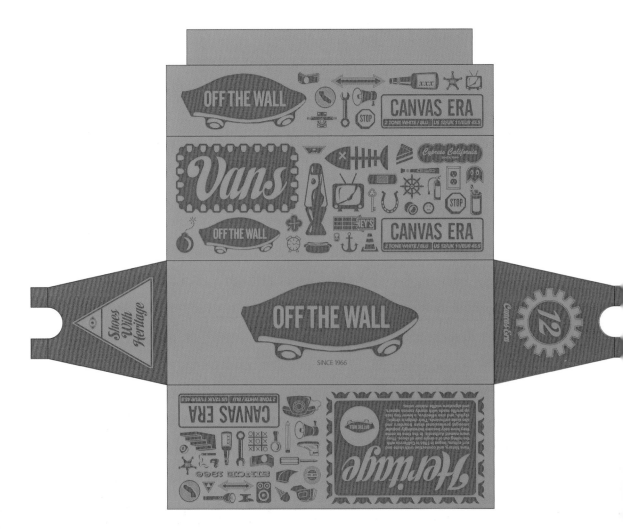

Five Ten Brand
Climbing Shoe Box

This is a two-part corrugated
box design. The pull out drawer
function gives a fresh take on the
typical shoebox.

Designer: Ryan Huettl
Production Date: 2011
Nationality: USA

Qubic Store Packaging

This year Qubic has embraced the 'recycle and reuse' philosophy, fitting out its store with a range of everyday items and materials used in engaging new ways. The two forms of packaging take their inspiration in part from the Japanese tradition of gift wrapping. 'The handle' has been specifically designed to accommodate garments of more or less any size. The purchase is folded inside selected wrapping paper and the corrugated handle is attached using self-adhesive panels. In conjunction with a second component the handle can be adapted to carry shoeboxes of various sizes. A carefully configured strip of card is folded around and over the carton and secure with the handle. This provides consistent branding whilst allowing the consumer to 'show off' their new purchase. The handle is perforated across the top, so the package can
be opened easily via ripping, similar to the traditional style of gift opening.

Design Agency: Casey Ng Studio
Production Date: 2011
Client: Qubic Store
Nationality: New Zealand

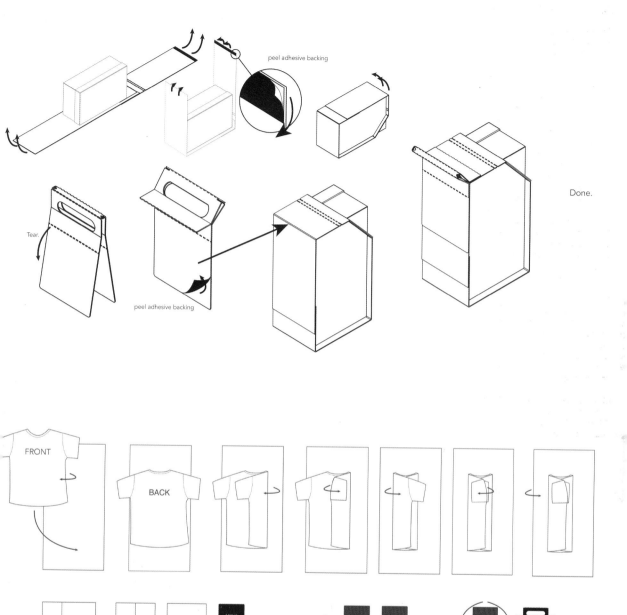

peel adhesive backing

Done.

Tear.

peel adhesive backing

FRONT

BACK

QUBIC STORE

Done.

Packaging Proposal for Fashion Brand

Using single-layer corrugated cardboard and coloured acetate film, the package, in two sizes and three colours, fits a wide range of products.

Design Agency: Enrique Romero de la Llana
Production Date: 2010
Client: Robe di Kappa
Nationality: Spain

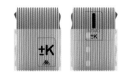

FONT: BRYANT
regular:
ABCDEFGHYJKLMNOPQRSTU-VWXYZ
bold:
ABCDEFGHYJKLMNOPQRSTU-VWXYZ

ETICHETTE AUTOADESIVE IN PLASTICA TRASPARENTE

299 EC   1788 EC   370 EC

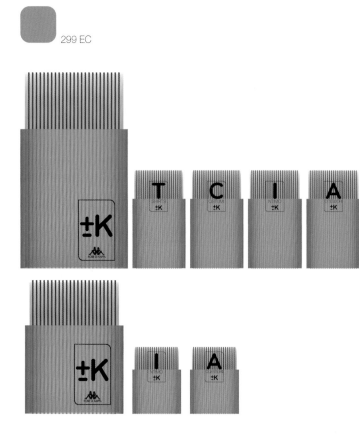

MAN

SPORT

WOMAN

T-SHIRTS

INTIMO

COSTUMI

ACCESSORI

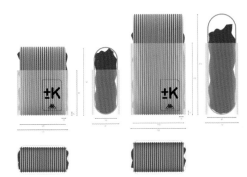

Air Jordan 2011

Getting back to the performance roots, a new modular midsole technology was introduced with the launch of our Air Jordan 2011, which required us to craft a product packaging narrative significantly deeper than a standard tech call out sheet. Nike worked with illustration collective Non-Format to develop a series of stylized information graphics articulating the functionality of the midsoles, which was executed on the sock liners themselves as well as a protective microfiber cloth in which the shoes came wrapped. The corrugated box was finished with a matte pp coating, tonal spot UV hits and a beveled area for the redesigned UPC label to sit.

Design Agency: Internal
Production Date: 2011
Client: Nike
Nationality: USA

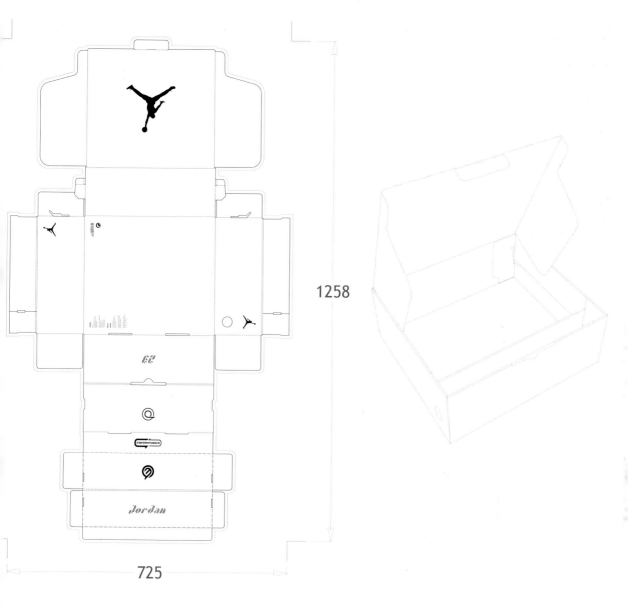

1258

725

Watchpackaging

In order to have an interesting and eco-friendly packaging solution for the watches, Steven Götz developed this product. It is made from two pieces of corrugated cardboard, that enclose the watch. The pack is light, doesn't take up much storage space and is easy to ship. In his opinion the mundane nature of cardboard brings out the quality of the Swiss watchmaking.

Design Agency: Steven Götz
Production Date: 2009
Nationality: Switzerland

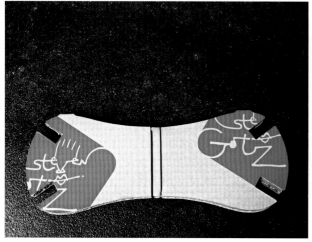

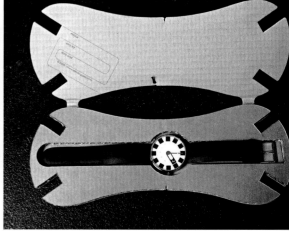

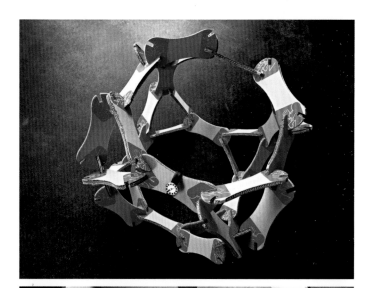

Nooka

Here designers wanted to design something so it didn't looked designed, they just wanted the watch to pop out and add a little extra futuristic touch to the packaging concept. Designers kept the package really clean and just added a special film made in Japan, that has the ability to blur in different angles. The choice of corrugated material adds to the simplicity of the design.

Designer: Maja Lehman, Madelene Hansson and Richard Feldéus
Production Date: 2010
Client: NOOKA
Nationality: Sweden

Giftbox : LINDE No1

LINDE No1, a Danish Design telephone - slim & simple - yet functional & fabulous. The corrugated giftbox is designed to appeal consumers who like a minimalistic lifestyle and modern design. The phone is eco-friendly and requires no batteries.

Design Agency: SOKAN telecom
Production Date: 2011
Client: SOKAN telecom
Nationality: Denmark

LINDE No 1

21 Drops

The goal of the design was to achieve a contemporary sensibility while respecting the heritage of aromatherapy and the artisanal nature of the product. This was achieved through the amalgamation of a vibrant colour palette and font-driven numerical graphics combined with embossed patterns and a corrugated.

Design Agency: Purpose-Built
Production Date: 2010
Client: 21 Drops
Nationality: USA

Spiezia Redesign

100% organic with strong ethical values, Spiezia wanted to marry style with substance in their re-branding. The natural qualities and Cornish provenance of the brand is conveyed via livery inspired by the St Ives School of Artists. Innovative use of corrugated materials, structural design and production have made the brand more sustainable, as well as more appealing to the eye.

Design Agency: jkr
Production Date: 2008
Client: Spiezia
Nationality: UK

Pangea Organics Skincare Discovery Kits

Pangea Organics Skincare Discovery Kits with travel size products enclosed. Three kits by skin type distinction. Each corrugated box displays the contents of each kit on the front flap of the package.

Design Agency: Pangea Organics In-House
Production Date: 2011
Client: Pangea Organics
Nationality: USA

DISCOVER YOUTHFUL RADIANCE IN 5 SIMPLE STEPS

SKINCARE DISCOVERY KIT
for normal to combination skin

ANTIOXIDANT-RICH FORMULAS. ANTI-AGING RESULTS.
BONUS EYE CREAM SAMPLE INSIDE

DISCOVER YOUTHFUL RADIANCE IN 5 SIMPLE STEPS

SKINCARE DISCOVERY KIT
for normal to dry skin

ANTIOXIDANT-RICH FORMULAS. ANTI-AGING RESULTS.
BONUS EYE CREAM SAMPLE INSIDE

Eco-friendly Package Design

This package and bottle design was devised to be fully recyclable. Recycled corrugated cardboard was die cut and layered with an eco- friendly adhesive in such a way to ensure a strong structure and sustainability, this also gave a textural element to the design and limited the need for printing and the use of inks.

Design Agency: Denton/Design
Production Date: 2010
Nationality: Australia

KOPIA/Modular Tableware

Kopia is a modular tableware family made in porcelain and cork, inspired by an ancient Hungarian ritual of 18th century: The Kopjafa was a wooden obelisk made not only to celebrate a funeral ceremony but to commemorate the life of a departed person as well, through geometric symbols that used to describe the sex, the character, the profession, the age (etc.) of such person. Corrugated package offers better protection of the tableware and an illustration of the design concept.

Designer: István Böjte
Production Date: 2012
Nationality: Hungary

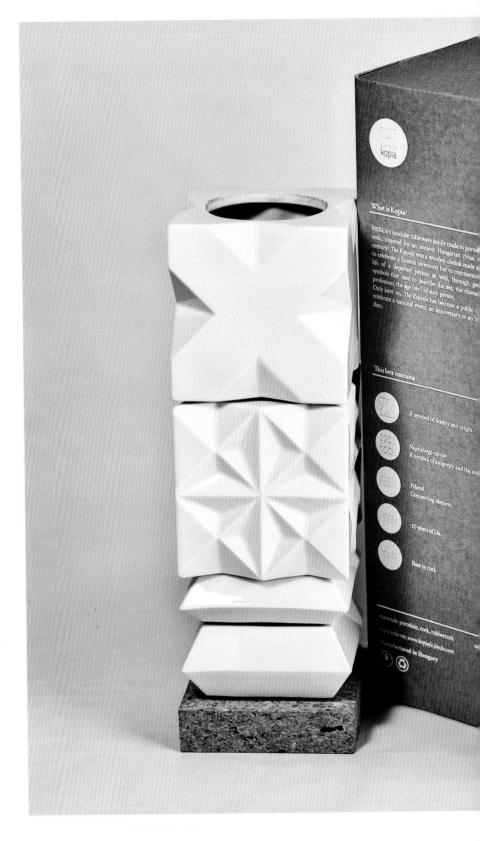

Skeleton Packaging

Skeleton Packaging is a conceptual design focused on producing sustainable packaging solutions for fragile objects. This goal is achieved by removing all excess paneling from the corrugated package design, leaving only the remaining 'skeleton' which is still fully capable of protecting the product from bumps and shakes. The overall structural integrity of the packaging is kept intact. Taking this recent spike of waste production into consideration, Skeleton Packaging is designed to use the least amount of material for manufacturing and consume less energy during its production process, resulting in a product that weighs less and is easily stacked for efficient transport.

Designer: Roy Sherizly
Production Date: 2011
Nationality: Israel

SKELETON PACKAGING

Libbey Shot Glasses

The objective was to redesign and improve sustainability on Libbey shot glasses package, which is currently outdated and unsustainable. The brand targets both males and females, youthful and active, between 18-35 years of age in middle income. The redesigned package was created on one piece of self-locking corrugated sheet. More than 50% of products can be revealed in the package to increase product visibility.

Designer: Jing Li, Rong Zhao
Production Date: 2011
Client: Libbey Inc.
Nationality: China

Carillon

The eloquent and simple corrugated packaging provides sturdy protection with peep holes to see the glasses placed inside. The glasses are protected from the other with panel slips that also serve as recipe cards. With glasses and recipes ready, all that is left to grab are some ingredients and ice and a friendly reminder to shake and not stir.

Designer: Michelle Tieu
Production Date: 2010
Nationality:USA

Orrefors Divine

Input

It was Orrefors, and designer Erika Lagerbielke, who were given the great pleasure of producing the official gift from the people to the Swedish Royal Wedding couple. This exclusive royal set of glasses is not available for the public, but a very similar 'retail version' was created and named 'Divine'.

Output

With the design and communication concept 'Love is Divine' the parallel to the royal set of glasses is rather obvious, as long as the Royal Wedding is on the agenda. When, however, all memories of the wedding have declined, the concept lives on and could easily be transformed into almost anything. Always with an appropriate illustration. Neumeister delivered the main concept, packaging design, ads, sales material ; brochure, point of sale material, etc. In terms of packaging, corrugated cardboard is chosen to provide absolute protection during transport and display.

Design Agency: Neumeister Strategic Design AB
Production Date: 2010
Client: Orrefors
Nationality:Sweden

Sustainable Thinking

O zone worked alongside Media Profili in the planning and realization of its corporate Christmas gifts for 2011. Following a sustainable approach, the designers came up with the project's basic elements: an aluminum SIGG bottle and a protected Slowfood food kit. The idea of the bottle is to consume less plastic and promote the reuse of a product. Rice is one of the territory's specialties, which Chef Matteo Caleffi combined with various other ingredients to come up with two recipes that the designers then illustrated and included in the packaging. Honeycomb corrugated cardboard, a natural and recyclable product, was decorated with custom Christmas graphics, yet still in line with the Media Profili image.

Design Agency: o zone
Production Date: 2011
Client: Media Profili S.p.A.
Nationality: Italy

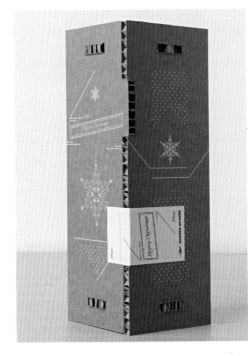

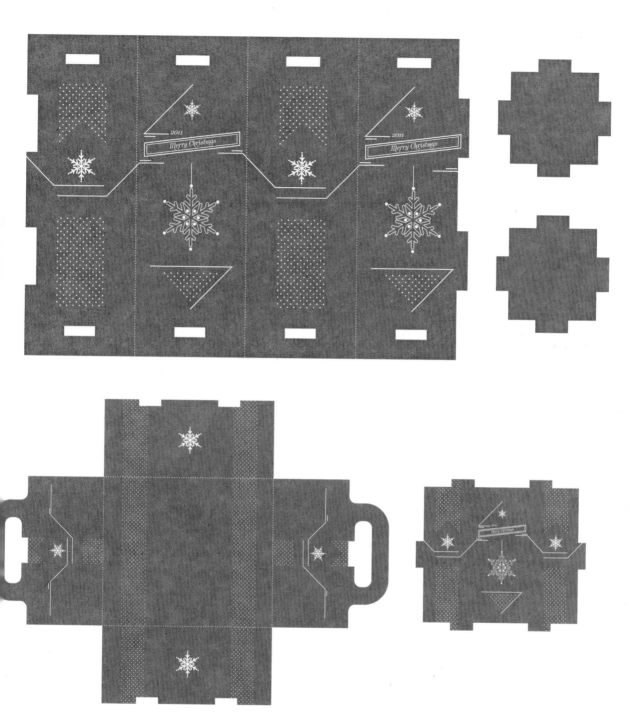

Sussner Design Company Materials

Sussner Design Co's identity is the hand - a palm with four flexible fingers and one opposable thumb. Sussner Design Company enjoy creating well-crafted self-promotional materials for ourselves including posters, pint glasses, brochures, coffee cups, office interior graphics and self promos, such as The Lunch Spinner. The package of corrugated material provides solid protection.

Designer Agency: Sussner Design Company
Production Date: 2009
Client: Sussner Design Company
Nationality: USA

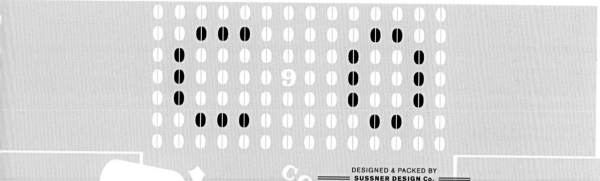

DESIGNED & PACKED BY
SUSSNER DESIGN Co.
MINNEAPOLIS, MN 55401
612.339.2886 | SUSSNER.COM
PRINTED BY
Steady
PRINT SHOP CO.

6 NO.
SURVIVAL ITEM

SD C⊙FFEE
BRAND

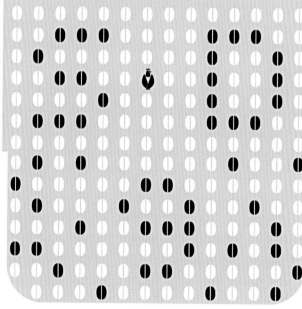

Sony Eco-Friendly Package

An eco-friendly approach was applied in the re-design of the Sony Cyber-Shot packages. It utilizes recycled paper, corrugated card and one colour printing. The re-design also uses 30% less material then the original, having a huge impact on cutting the cost of shipping, transportation and storage; which can hugely affect the market price.

Designer: Lisa Lauren Tersigni
Production Date: 2010
Nationality: Canada

Moon

For the package design, the designer applied a silk-screen printing to corrugated cardboard, since they purposely created this to be simple.

Designer: Hidetoshi Kuranari, Chihiro Konno
Production Date: 2010
Nationality: Japan

Light Bulb Package

The aim was to create a design concept for a product, using corrugated cardboard as a material. The cardboard is made of fragile material (paper) but it is commonly used in a packaging to protect products. The designer chose the light bulb for her product which has a similar characteristic. The packaging for a light bulb is not only protective, but also the design pattern and the twisting motion while the person opens the packaging creates exciting and dynamic tension of paper. The colours were chosen to match with the idea of brightness / darkness of the light.

Designer: Soo Yeon Park
Production Date: 2011
Nationality: Korea

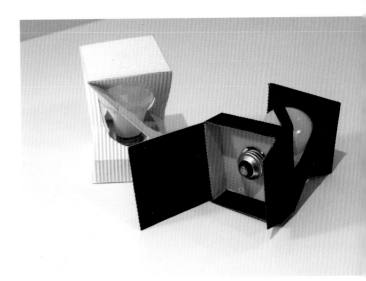

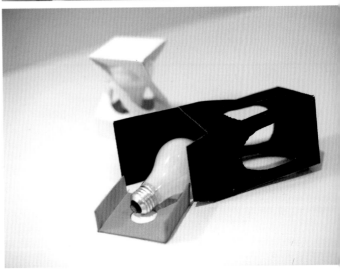

Lunar Light Bulb Packaging

Sustainable packaging for energy-efficient light bulb based on the concept of the lunar eclipse. The lunar eclipse theme is reflected in the colours, texture and structure. The packaging is created using four pieces of corrugated cardboard glued together and the label is printed on the cardboard surface with soy ink, minimizing waste and promoting sustainability. The inner structure, shaped with a single piece of cardboard, provides complete protection for the fragile glass bulbs while the exterior packaging makes a transparent display.

Designer: Angel Jo
Production Date: 2009
Client: Lunar light bulb
Nationality: USA

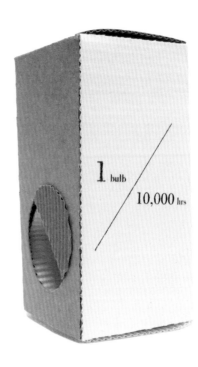

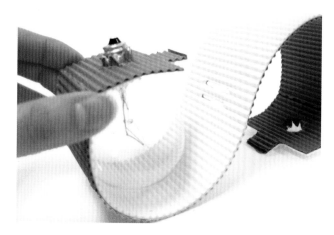

Round Bulb Packaging

The following packaging solution seeks to minimize the amount of material necessary to protect a set of three 60 watt round bulbs. Inspired by the fineness of Japanese egg packaging, the kraft corrugated paper gracefully receives each bulb, creating a special unwrapping experience and a rather sculptural object. The cardboard is sturdy and 100% recyclable and offers the chance to rip the structure as it is used.

Designer: Oscar Salguero
Production Date: 2010
Nationality: Peru

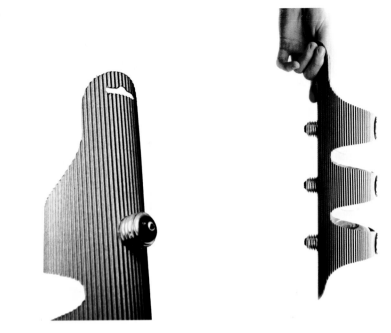

299.72 mm

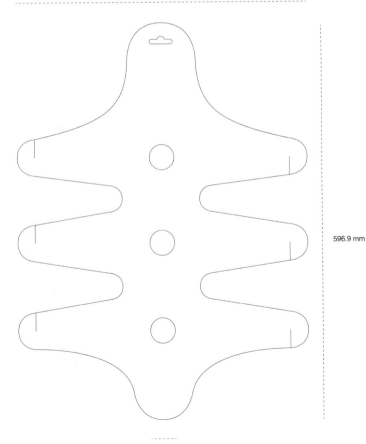

596.9 mm

25.4 mm

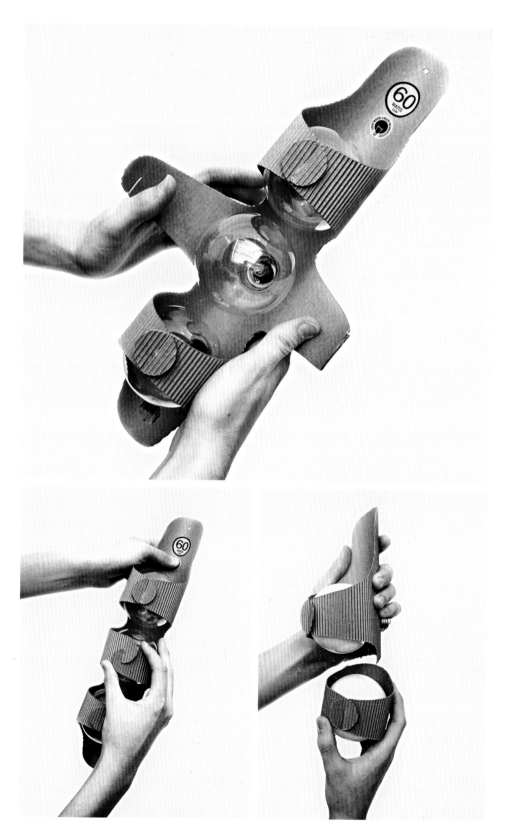

Boost Mobile
Replacement Phone

Simple branding and typography treatments to a craft box allow for the customer to quickly identify the corrugated package within the mail and get back up and running with their new prepaid mobile phone.

Design Agency: Bay Cities Container
Production Date: 2009
Client: Boost Mobile USA – Sprint.
Nationality: USA

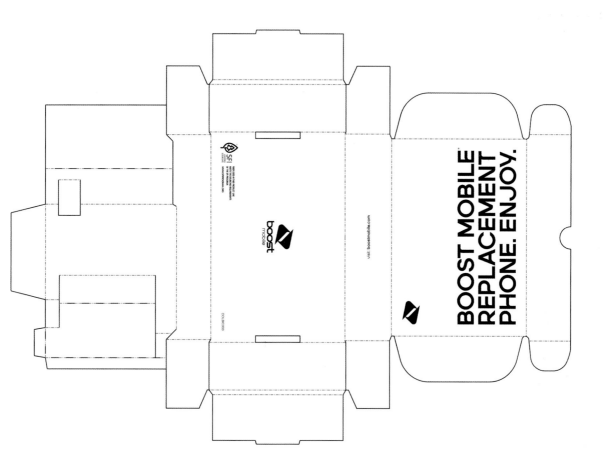

Bajaj Electricals

Leo Burnett designed a corrugated packaging unit which would not merely be a storage space for the iron, but would offer customers an added utility in the process of ironing. Thus developed the special edition Fold Aide Box. The packaging unit not only stores the iron but also assists in the process of folding ironed clothes.

Design Agency: Leo Burnett, Mumbai
Production Date: 2010
Client: Bajaj Electricals
Nationality: India

The Fold Aide Mechanics

1. The ironed shirt is placed face down on the open box as shown.

2. The bottom part of the shirt is folded up till a marked line.

3. The left panel of the box is folded as shown.

4. The right panel is now folded.

5. The left panel is then folded once again.

6. Next the bottom panel is folded.

7. The ironed shirt is now folded perfectly.

8. The end result.

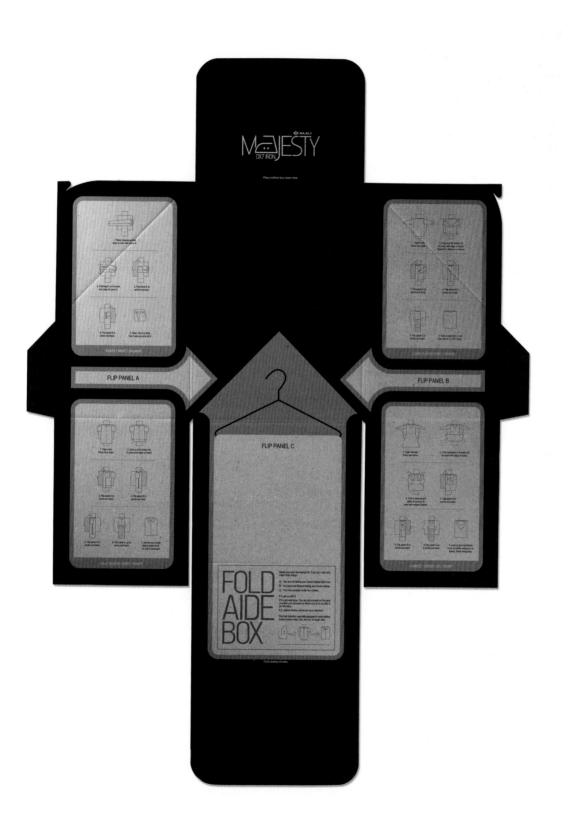

'Three to Six' is based on a simple and clean organic chair design. Its inspiration comes from high-end chair like Dutch's 'Vouwwow' and Eames's 'Zig Zag'. The designer wants to transform a recycled material like corrugated cardboard into a useful daily object--a chair. This transformation is a product of what was once a horizon of possibilities and now a product of beauty and elegance.

Design Agency: BiolArt
Designer: Sander Jackson Siswojo
Production Date: 2010
Client: ID3
Nationality: Indonesia

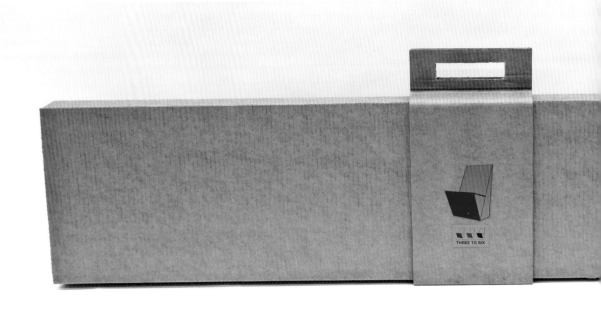

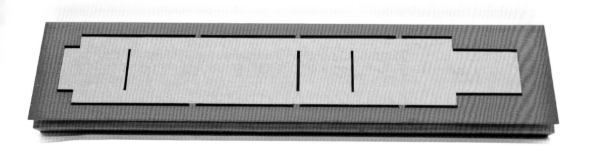

Twist

Not many people use loofah sponges for cleaning, which is why Twist is getting a rebrand. Now with a more organic and feminine appeal, more women will be attracted to using an eco-friendly solution to cleaning. Included is a brand extension of loofah sponges for the body. And corrugated cardboard is chosen for the package since it is easy to cut and craft.

Designer: Boya Liu
Production Date: 2012
Client: Twist rebrand
Nationality: USA

Anydesign Welcome Kit

Agency's Welcome Kit for Gloria Kalil's Fashion Marketing 2008. The mini corrugated sewing box contains the prospectus of the agency, developing an analogy with a tailor.

Design Agency: Anydesign (Brazil)
Production Date: 2008
Nationality: Brazil

Provenance Packaging

Provenance makes high-quality homeware products from recycled, reclaimed and renewable materials. In keeping with the products, the packaging is designed to achieve high shelf and low environmental impact. Strong orange is used for the boxes to draw attention to the display, the corrugated board is 100% recycled, and is left unbranded to encourage reuse. It is self-coloured to make any in-store damage less visible, reducing the need for re-boxing. Branding is restricted to the paper sleeves to minimize waste when adapting packaging to different languages.

Design Agency: Jog Ltd.
Production Date: 2011
Client: Provenance
Nationality: UK

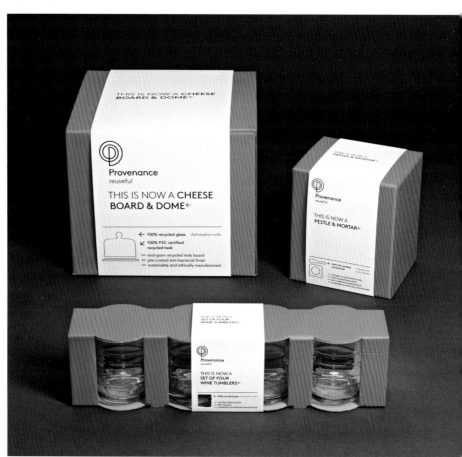

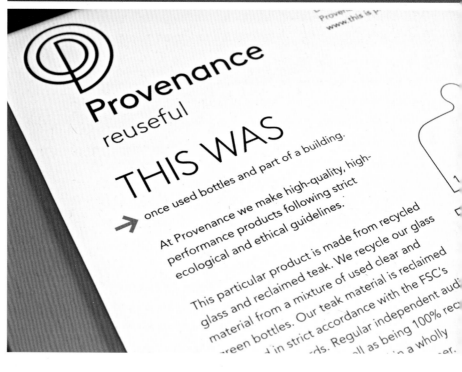

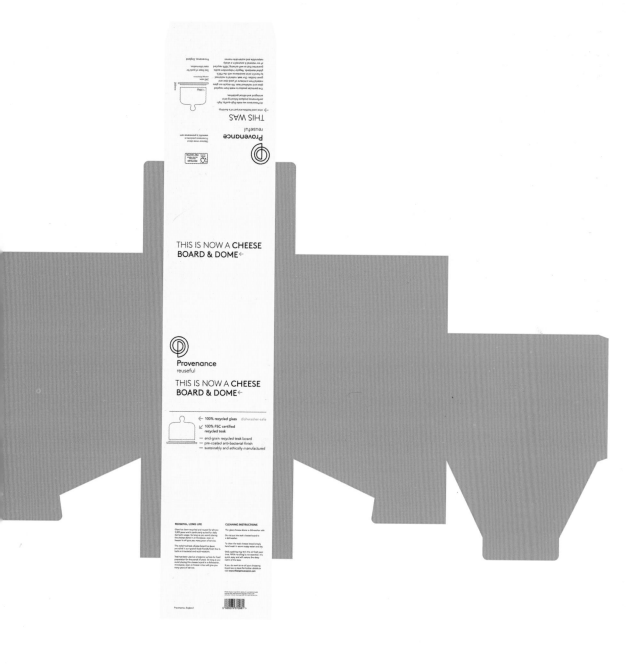

THIS IS NOW A **CHEESE BOARD & DOME**←

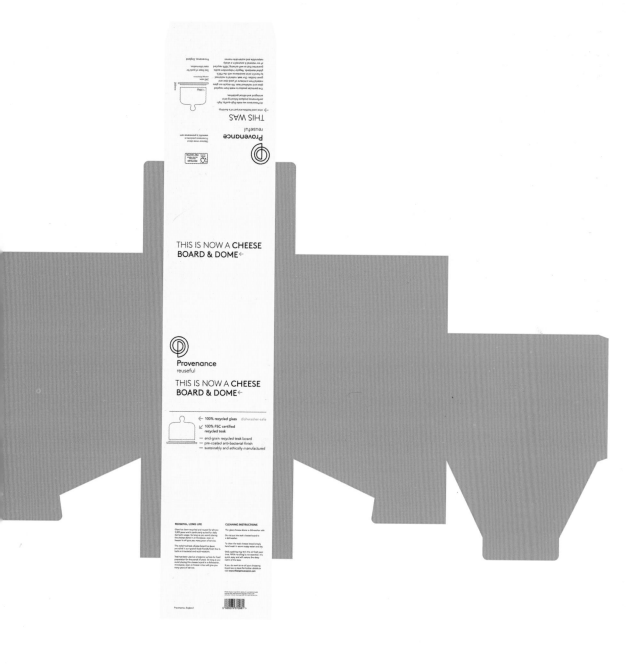

Provenance
reuseful

THIS IS NOW A **CHEESE BOARD & DOME**←

← 100% recycled glass   dishwasher-safe

↙ 100% FSC certified
    recycled teak

— end-grain recycled teak board
— pre-coated anti-bacterial finish
— sustainably and ethically manufactured

REUSEFUL, LONG LIFE

Glass has been recycled and reused for almost 3,000 years and is particularly suited for daily domestic usage. So long as you avoid storing the cheese dome in a microwave, oven or freezer it will give you many years of service.

The reclaimed teak cheese board has been pre-oiled in our special food-friendly finish that is both anti-bacterial and stain-resistant.

Teak has been used as a hygienic surface for food preparation for thousands of years. So long as you avoid placing this cheese board in a dishwasher, microwave, oven or freezer it too will give you many years of service.

CLEANING INSTRUCTIONS

The glass cheese dome is dishwasher safe.

Do not put the teak cheese board in a dishwasher.

To clean the teak cheese board simply hand wash in warm soapy water and dry.

Daily washing may thin the oil finish over time. While re-oiling is not essential, it is quick, easy and will restore the deep lustre of the teak.

If you do want to re-oil your chopping board we share this item in store. For further details or visit www.thisisprovenance.com

THIS WAS

Provenance
reuseful

THIS WAS →
once used bottles and jars of a bakery.

Babiloon

Reduce consumption. Limit waste. Recycle. Give back. Every day in the office gives you an opportunity to eliminate habits that are bad for the environment. These five corrugated cardboard foundations will brightly encourage you to do so.

Designer: Urska Hocevar
Production Date: 2010
Client: Kaaita
Nationality: Slovene

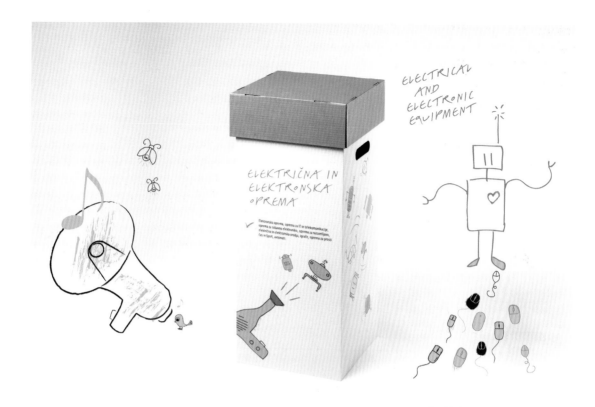

ELECTRICAL
AND
ELECTRONIC
EQUIPMENT

ELEKTRIČNA IN
ELEKTRONSKA
OPREMA

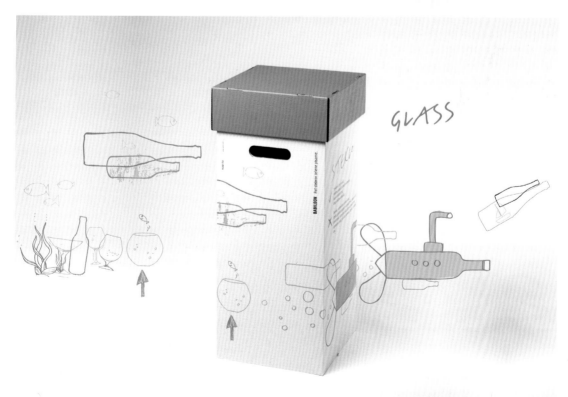

GLASS

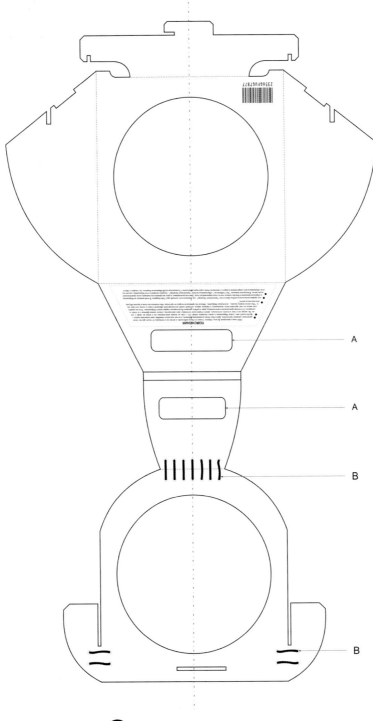

A Laces

B A groove for a hand

The Spirit of Football

The spatial sculpture of the corrugated packing perfectly works as the stand and the attention in shop will draw, unlike boxes. Thus packing is convenient for transferring. At moving the packing well works as advertising! And its sand colour associates with Africa! At packing there is an irreplaceable element of football equipment-a lacing. The lacing theme can be developed and as an identification element to have colours.

Designer: Ilya Avakov
Production Date: 2009
Nationality: Russia

X-Acto

DIY meets utilitarian. The reusable X-ACTO packaging recognizes the user's every need in terms of transport, storage, usage, and disposal. Additionally, the compartmentalized design ensures that all of your precision tools are safely and conveniently accessible. Packaging looks as if it could have actually been cut out with an X-ACTO blade. The stacked design utilizes B-fluted corrugated and soy-based flexographic inks, making it lightweight, durable, and sustainable. With a cutting-edge design that satisfies both form and function, the new X-ACTO packaging is the all-in-one solution for every designer and artist.

Designer: Joy Lin
Production Date: 2010
Client: X-Acto
Nationality: USA

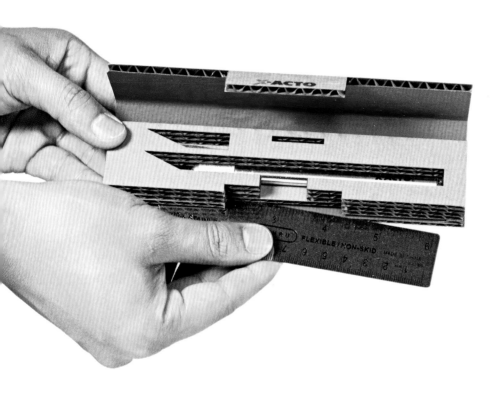

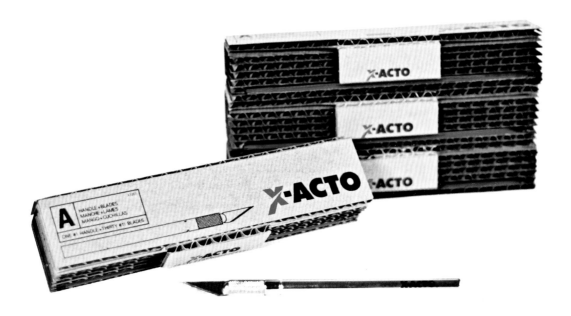

Fix It

FIX IT is a tool box for the young adults, that have just moved to their own home and don´t have a proper collection of tools. The designer is not trying to replace the professional tool box with FIX IT. The idea was to create ecofriendly, inexpensive, organized and good-looking tool box for everyone. Usually these toolboxes are made of stainless steel, wood or plastic. The designer wanted to try if it´s possible to create it out of corrugated cardboard. The brief was to design a cardboard packaging for a new target, better solution for an old product or totally new and innovative way of using the material.

Designer: Anu Nokua
Production Date: 2011
Nationality: Finland

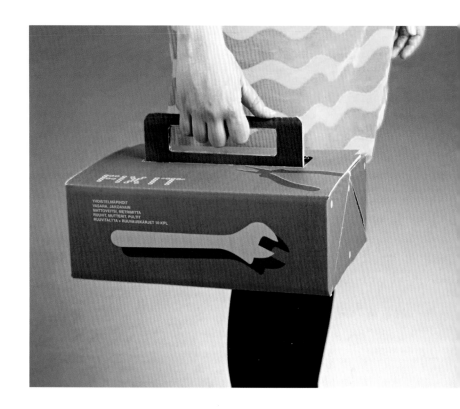

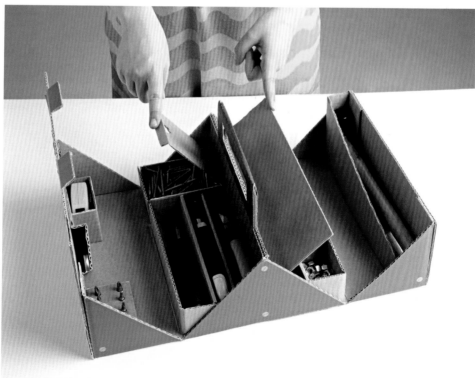

AUTA / Help

First aid kit made of corrugated cardboard. AUTA is a first aid kit. It includes antiseptic wipes, band aids and gauzes. The packaging is easy to open and carry because of the handy lid. The graphics of AUTA will show you what is inside of it. The idea was to use the structure of the cardboard liner as a graphical element in the packaging.

Designer: Anu Nokua
Production Date: 2011
Nationality: Finland

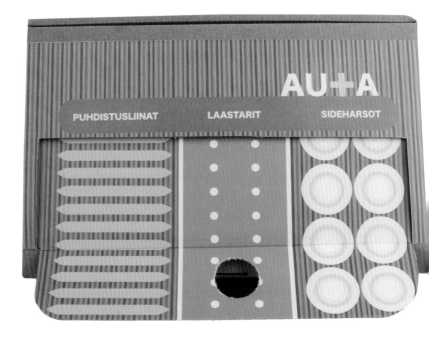

Sprout Family

Sprout Family is an indoor garden starter kit for kids who don't have an opportunity to enjoy nature. By growing plants, kids would naturally develop their responsibility to love living things. Made of corrugated carton: These are safe corrugated boxes to protect seeds from breaking and moisture. Also, it is 100% recyclable.

Design Agency: So Young Studio
Production Date: 2010
Nationality: Korea

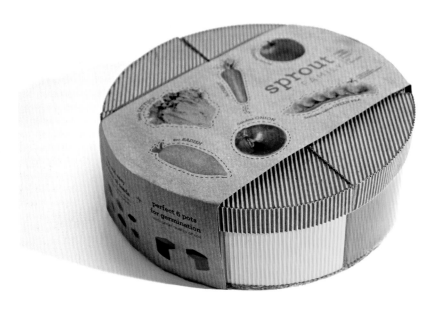

Jobeur - Nail Packagaging

The problem was to design an eco-friendly packaging with the primary function of storing multiple objects. The proposed solution requires a single rectangle of corrugated cardboard and uses no ink and no glue point. The packaging capitalize on the head of nails to indicate the sizes and shapes of the product.

Designer: Pier-Philippe Rioux
Production Date: 2011
Nationality: Canada

De Fietsfabriek

'De Fietsfabriek' is translated
'The Bike factory' in english. It's
a Dutch bicycle company that
manufacturers and sell handbuild
quality bikes and the assignment
was to redesign their existing
brand. The approach were to
unite the robust bikes and the
handcraftet production without
loosing the product. The labelling
and printed images outside
the corrugated package help
distinguish the products from one
another.

Designer: Kristoffer Hvitfeldt
Production Date: 2012
Client: De Fietsfabriek
Nationality: Denmark

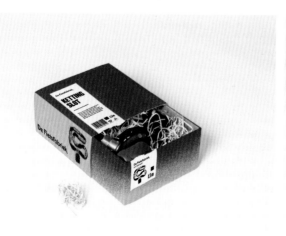

Selle San Marco
Vintage 1935 Line

The Vintage 1935 line of Selle San Marco is the collection of the products that made the history of cycling in the 70s and 80s. The aim of the corrugated package is to provide the same feeling and emotion of glorious past, handmade materials and the Italian flavour.

Design Agency: Quasidesigner. com
Production Date: 2008
Client: Selle San Marco S.p.A.
Nationality: Italy

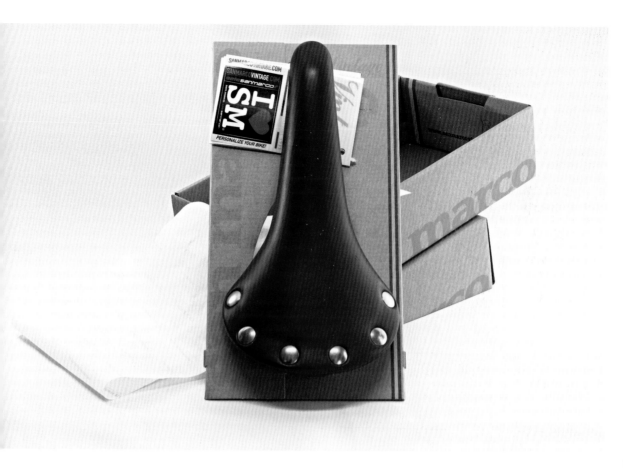

Fume

A packaging and product design project, made up of re-used materials. A hand crafted smooth edge wood tobacco pipe, whose identity reflects a modern individual. The package design consists of a layered corrugated cardboard with a simple geometric form, providing space for the product.

Design Agency: I AM HANNAH
MATTHEW DESIGNS
Production Date: 2012
Nationality: USA

Leafy Dream

The idea of this project was to design a package for a jewellery tree. Main concept was to make a person, who is holding the package, feel familiar with the Jewellery Tree, that is inside it. Smile Group decided to choose an ecological and simple structure of the package, as it saves the energy and time needed to produce it. Also, Smile Group wanted to make it a little bit interactive, so person discovers the 'content' little by little, by performing opening actions. A great interest was paid to make this micro-corrugated package suitable for post delivery, so it is strong and reliable and still meets the smallest sending requirements.

Design Agency: 'Smile Group'
Production Date: 2012
Client: jauku | design
Nationality: Lithuania

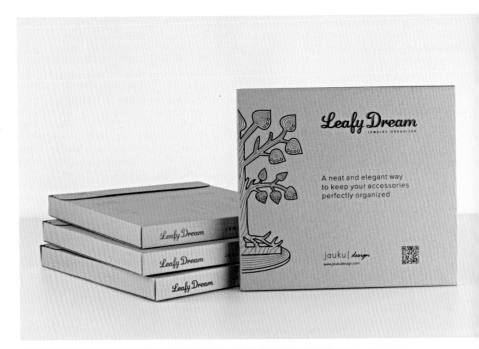

TobeUs Packaging Design

'Playing is a serious matter' as designer Bruno Munari used to say. Corrugated cardboard is used for the package, as its printability works well for such designs with a large amount of design information to present.

Design Agency: Due mani non bastano
Production Date: 2010
Client: TobeUs
Nationality: Italy

Calendar and Packaging

This perpetual calendar and strong corrugated packaging at the same time, can be used over and over again, every following year, and contains a bag made of recycled material and a notebook with some illustrated information about types of wastes that we deal with in everyday life.

Design Agency: Kitchen&GoodWolf
Production Date: 2010
Client: City Administration for Environmental
Nationality: Serbia

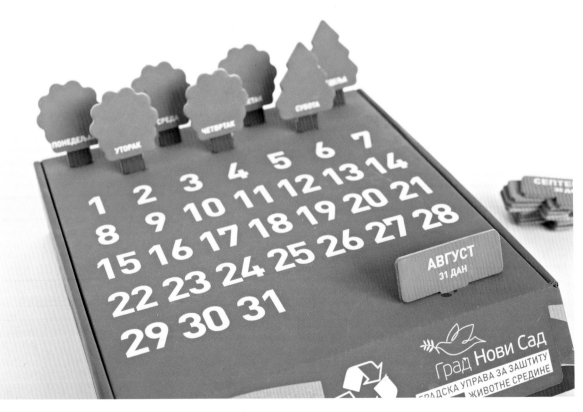

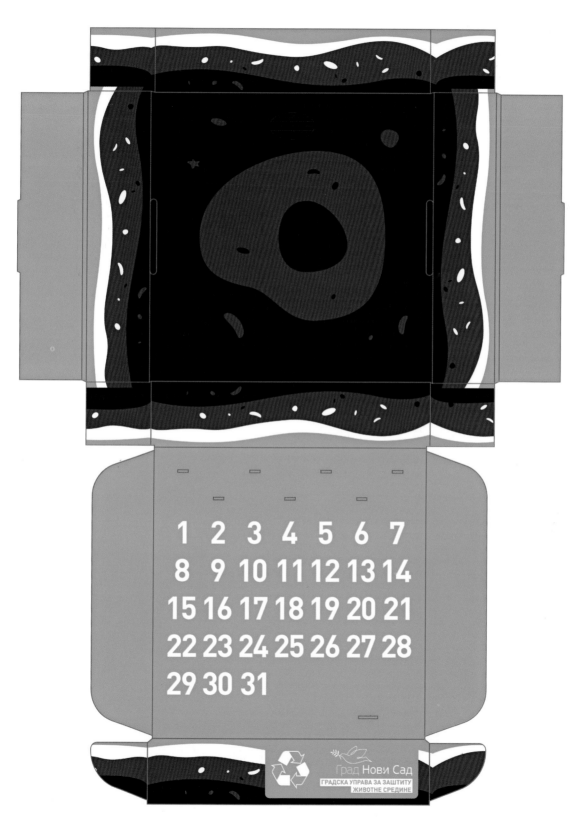

Pure Coal

The barbecue's tools are made from recycled aluminum and reclaimed, recycled, and FSC woods. Moreover, committed to inspiring a fundamental shift toward environmental responsibility in the food industry, the line Pure Coal® by Kingsford are packaged with only sustainable materials, mainly with corrugated cardboard.

Design Agency: Winnie Yuen
Production Date: 2011
Nationality: China

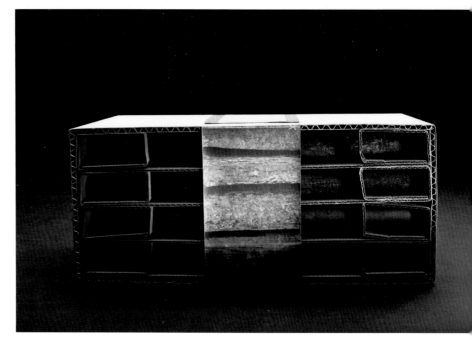

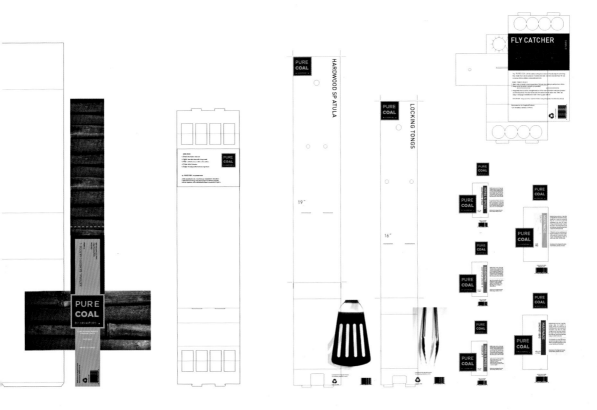

Oi Talking Sofa

Sofas and other seating arrangements are sold and shipped usually in one piece. This carries a high cost to ship and move most pieces not to mention the environmental impact of shipping large heavy single units. The Oi seating pods are created to approximately three times the amount of seating than traditional sofas. These modular pods can create an infinite amount of configurations by adding pods together. They are also able to be moved by one person, easily passing through doorways and around corners.

Designer Agency: Cocoon
Production Date: 2010
Client: Oi Funiture
Nationality: Canada

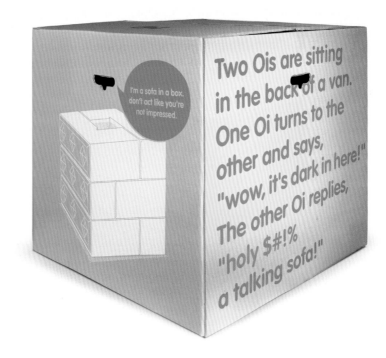

Light Bulb Package

A quirky design for a serious product. This eco-friendly corrugated packaging for environmental light bulbs speaks to you from the shelves with phrases that will catch your eye.

Designer: Leilani Silversten
Production Date: 2008
Nationality: USA

Macal Blankets, Presentation Kit

The packaging/concept solution was a white card box with the concept in one side 'WE CARE ABOUT THE QUALITY OF YOUR DREAMS' and the images on the other, with the logo on the top. When you open it, you have 10 spines with different colours and the product identified as dream one, dream two...

Design Agency: António Queirós Design
Production Date: 2012
Nationality: Portugal

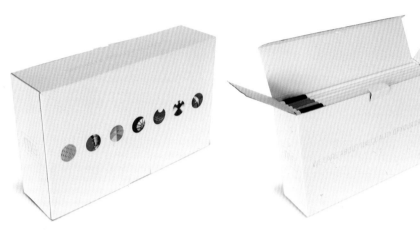

XO Packaging

XO aims to keep design, technology and user experience balanced through their natural handwriting solution. The packaging is made up of the elements that are clean, simple and natural ; which suits the design philosophy.

Design Agency: cloudandco
Production Date: 2011
Client: XO (Byzero Inc.)
Nationality: Korea

Imani Kimpe Birthday Card Packaging

For the birth of the designer's first-born child Imani, he really wanted an original birthday announcement, as well by idea & concept as in use of material and execution of the work. The designer wanted to use unconventional materials, which meant it probably wouldn't be simple to get his hands on it and keeping it budgetwise affordable. So he chose corrugated cardboard for the birthday card which is both recyclable and inexpensive.

Designer: Gotcha!
Production Date: 2009
Nationality: Belgium

Imani birthday card

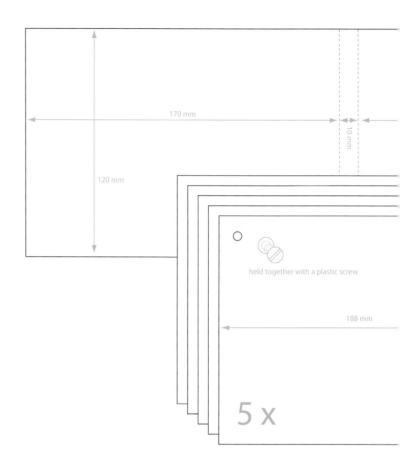

170 mm

10 mm

120 mm

held together with a plastic screw

188 mm

5 x

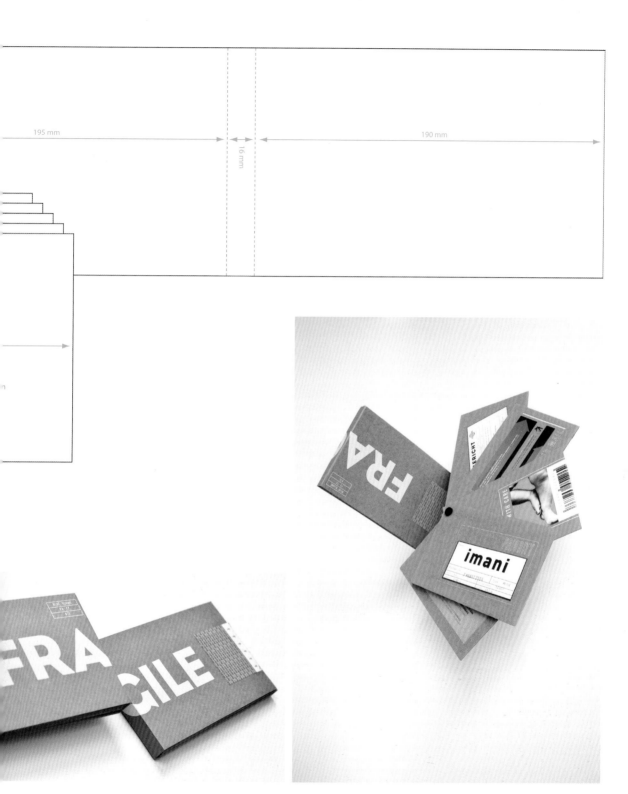

195 mm

16 mm

190 mm

Pencil Block II

A convenient desktop organiser constructed from laser-cut corrugated cardboard profiles. Through the re-implementation of this waste material, with a manufacturing process which manipulates the material rather than recycling the material, a usable product is created which effectively solves the problem of a messy desktop.

Design Agency: Jozi Design
Production Date: 2011
Nationality: South Africa

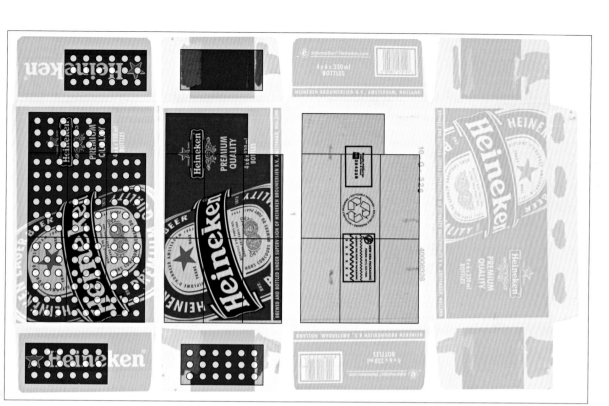

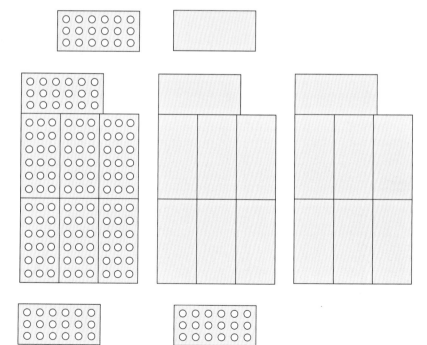

Self Promotion Mailer

This is a personal project to promote the designer himself as a designer and try to get some attention from design agencies. Gary Corr has the opportunity to gain some practical experience within the design industry, so rather than sending a covering email and a CV, he designed and produced this hand-made corrugated cardboard direct mailer to send out to potential employers, it contains links to Gary Corr online portfolio as well as other bits and pieces.

Designer: Gary Corr
Production Date: 2012
Nationality: Ireland

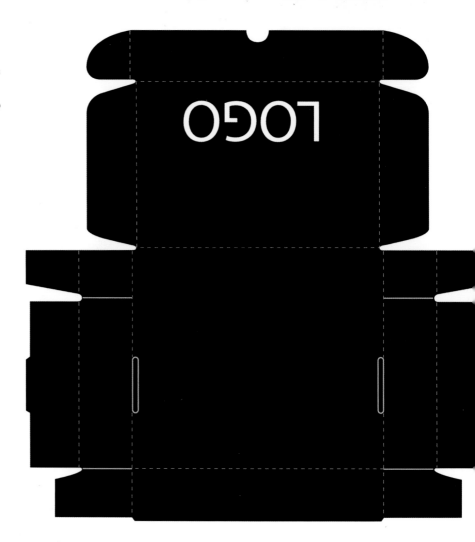

'Apocalypse: Emergency Boxes'

A collection of mass produced corrugated fiberboard boxes intended to be filled with flyers and emergency information in times of catastrophe. The project injects humour into information and facts that are usually used by the popular media as a tool to invoke hysteria.

Design Agency: GLD/FRD
Production Date: 2011
Nationality: Israel

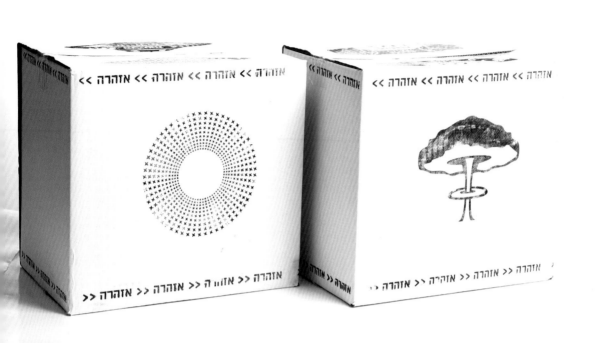

Hua-Shiang

The box Hua-Shaing is designed for sending back and forth for the purpose of communication, so the safety and durability is the primary consideration. The quality of corrugated paper and the structure of the package secure all objects inside the box against external hit while sending it.

Designer: KuangTa Yang
Production Date: 2011
Nationality: Taiwan, China

Eco & UD House

Design of packaging made
of corrugated cardboard
containing the Eco & UD House's
presentation of Panasonic, an
actual and overall proposal
of property to be reality in
Tokyo in 2010, that combines
the environmental respect, the
energy efficiency and the facility
of use of systems and products
that integrates.

Design Agency: uauh!
comunicación gráfica
Production Date: 2007
Client: Panasonic.
Nationality: Spain

Producción estuche: Pack especial

PIEZA 1
cara exterior / color blanco

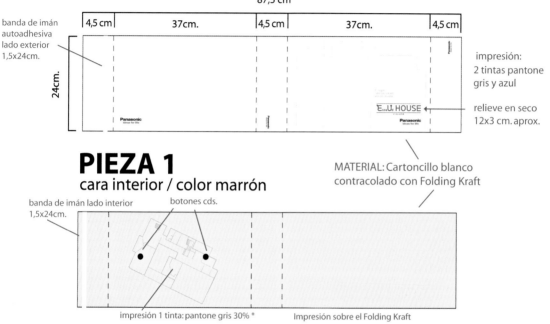

banda de imán
autoadhesiva
lado exterior
1,5x24cm.

87,5 cm

| 4,5 cm | 37cm. | 4,5 cm | 37cm. | 4,5 cm |

24cm.

impresión:
2 tintas pantone
gris y azul

relieve en seco
12x3 cm. aprox.

E...U. HOUSE

Panasonic

PIEZA 1
cara interior / color marrón

banda de imán lado interior
1,5x24cm.

botones cds.

MATERIAL: Cartoncillo blanco
contracolado con Folding Kraft

impresión 1 tinta: pantone gris 30% *

Impresión sobre el Folding Kraft

PIEZA 2

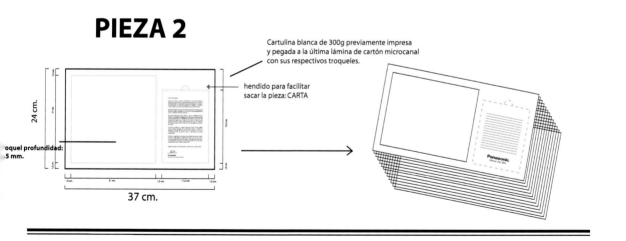

Cartulina blanca de 300g previamente impresa
y pegada a la última lámina de cartón microcanal
con sus respectivos troqueles.

hendido para facilitar
sacar la pieza: CARTA

24 cm.

oquel profundidad:
.5 mm.

37 cm.

Panasonic

Ulli & Chris

Ulli and Chris Laine- Valentini got married and invited their family and friends to celebrate with them in a rustic Alpine setting. The international guests enjoyed a beautiful evening at the Bödele, a hill close to Dornbirn in Austria. The corrugated package included a gingerbread heart and hay in reference to the alpine ambience.

Design Agency: Sägenvier Design Kommunikatation
Production Date: 2009
Client: Ulli & Chris Laine Valentini
Nationality: Austria

I Protect

The client wanted to give the
book 'The Wounded Land' at
a conference on sustainability.
Same as corrugated cardboard
protects the planet, they decided
to protect the book making it
a corrugated box. In the sash
and the interior of the box can
be read: I protect. Production
method: Offset, engraving and
dry stamp.

Design Agency: Dosdecadatres
Production Date: 2008
Client: AFCO
Nationality: Spain

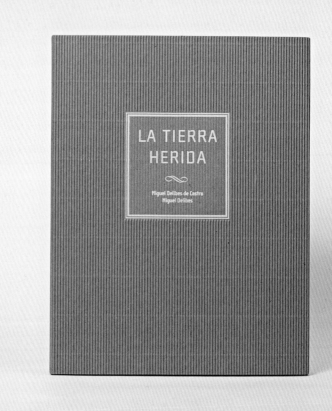

Sustainable DVD
Packaging for Nexus

For the Nexus Productions
DVD packs Beccy wanted to
create something much more
sustainable than the usual plastic
jewel case and came up with the
concept of a multi-functional,
100% recyclable corrugated
cardboard box, inspired by the
visual language and utilitarian
functionality of the humble
transit box. Collaborating with
the designers at Julia, Beccy
realised that vision with a
cleverly designed package using
traditional brown corrugated
cardboard with white screen-
printed type; that works not
only as a DVD case but also the
envelope, achieving the aim
of reducing waste while still
looking rather nice.

Designer: Julia (creatively
directed by Beccy McCray,
Nexus Productions)
Production Date: 2009
Client: Nexus Productions
Nationality: UK

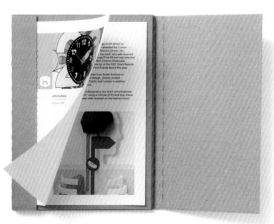

Index

Julia

kissmiklos

Konstantin Kolyubin,Lasha Kasoev

Kristof Kimpe

Kristoffer Hvitfeldt

KuangTa Yang

Leffe Goldstein Graphics

Leilani Silversten

Leo Burnett,Mumbai

Lisa Lauren Tersigni

Lluís Serra & Mireia

Luz Selenne Guardado

Maegan Brown

Maja Lehman, Madelene Hansson and Richard Feldéus

Michal Marko

Michelle Tieu

Nadia Arioui

Nate Eul

Neumeister Strategic Design AB

o zone

Olesya Kurulyuk

Olson

Olssøn Barbieri

Oscar Salguero

Pangea

Pier-Philippe Rioux

Purpose-Built

Quasidesigner.com

Roy Sherizly

Ryan Huettl

Sägenvier Design Kommunikatation

Sarah Machicado

Sarah Sabo

Say – Brand strategy & expression Productions

So Young Studio

SOKAN telecom

Soo Yeon Park

Steven Götz

Sussner Design Company

Tim Sumner

uauh!

Urska Hocevar

Winnie Yuen

Zoo Studio

© 2014 by Design Media Publishing Limited
This edition published in Apr. 2014

Design Media Publishing Limited
20/F Manulife Tower
169 Electric Rd, North Point
Hong Kong
Tel: 00852-28672587
Fax: 00852-25050411
E-mail: suisusie@gmail.com
www.designmediahk.com

Editing: Germán Úcar
Proofreading: Chen Zhang
Design/Layout: Muzi Guan

ISBN 978-988-12969-6-2

Printed in China